FERTILE GROUND:
NUTRIENT TRADING'S POTENTIAL TO COST-EFFECTIVELY IMPROVE WATER QUALITY

PAUL FAETH

WORLD RESOURCES INSTITUTE
WASHINGTON, DC
2000

ROBERT LIVERNASH
SENIOR EDITOR

HYACINTH BILLINGS
PRODUCTION MANAGER

AGRICULTURAL RESEARCH SERVICE, USDA
COVER PHOTO

Each World Resources Report represents a timely, scholarly treatment of a subject of public concern. WRI takes responsibility for choosing the study topics and guaranteeing its authors and researchers freedom of inquiry. It also solicits and responds to the guidance of advisory panels and expert reviewers. Unless otherwise stated, however, all the interpretation and findings set forth in WRI publications are those of the authors.

ISBN: 1-56973-197-7

Library of Congress Catalog Card No. 00-104473

Printed in the United States of America on chlorine-free paper with recycled content of 50%, 20% of which is post-consumer.

CONTENTS

FIGURES

TABLES

ACKNOWLEDGMENTS

I have many people to thank at the state agencies, on the advisory panels, and among my research partners in each of the three case studies covered in this study. From them, I have learned a great deal about the various issues involved in water quality and market-based mechanisms.

I owe special thanks to Dave Batchelor of the Michigan Department of Environmental Quality. Dave began life as a regulator, committed to improving the environment through enforcement of the law. While still maintaining that commitment, he is now working to improve environmental performance by identifying ways to appropriately increase flexibility and reduce costs. Dave has vast practical experience, which I found enormously valuable. In the course of this effort we spent many hours over the phone or a beer, thrashing through the issues, sometimes agreeing, sometimes not, but always working toward a better understanding and hopefully, better results.

Mahesh Podar at the U.S. Environmental Protection Agency also deserves a special mention. His office not only provided financial support for this project, but Mahesh led EPA's effort to explore trading in the nation's watersheds. That work helped to get a lot of people thinking about the possibilities and challenges and informed the analysis presented here.

In each of the three case studies, I worked closely with the lead state regulatory agency. We also formed advisory panels to review the work as it progressed. These panels provided invaluable critiques and helped to bring the analysis down to earth.

In Michigan, various individuals from several groups participated, including the Michigan Agricultural Stewardship Association, the Michigan State University Institute for Water Resources, MSU Extension Service, the Natural Resources Conservation Service, Consumers Energy Company, the Michigan Farm Bureau, the City of Wyoming, and the Huron River Watershed Council.

In Minnesota, lead partner Norman Senjem of the Pollution Control Agency was generous in sharing his insights with me. Our advisory panel in Minnesota included the Minnesota River Joint Powers Board, the Minnesota River Agriculture Team, the Natural Resources Conservation Service, the Minnesota Farm Bureau, the Metropolitan Council, the University of Minnesota, the Institute for Agriculture and Trade Policy, the Board of Soil and Water Resources, the Land Stewardship Project, and the Friends of the Minnesota River.

Our work in Wisconsin was completed in collaboration with the Wisconsin Department of

Natural Resources (WDNR) and the Rock River Watershed Partnership, a group of mostly point-source dischargers interested in cost-effective means of attaining clean water objectives. Sanjay Syal and Danielle Valvassori were the leads for the WDNR; Roger Sherman was the chair of the Rock River Partnership.

Two key partners helped complete the analysis. At the MSU Institute for Water Resources, Jon Bartholic and Da Ouyang provided hydrologic information needed to convert erosion and runoff estimates into loadings. One particularly vexing problem was solved by Mike Doran of Strand Associates, who developed up-to-date cost curves for phosphorus control technologies.

Thanks go to those who participated in the peer review process: Will Anthony, Dave Batchelor, Rob Day, Debbie Farmer, Al Hammond, Tony Janetos, Richard Kashmanian, Jonathan Lash, Mahesh Podar, and Marc Ribaudo. Bob Livernash provided review, editing, and production help. Kim Kusek and Andi Thomas deserve thanks for providing programmatic assistance. Hyacinth Billings, as ever, did a stellar job of seeing the report through production.

No effort can be accomplished without funding, and here a special acknowledgment is reserved for the Joyce Foundation, the Kellogg Foundation, the McKnight Foundation, the Oak Foundation, and the U.S. Environmental Protection Agency.

P.F.

FOREWORD

After nearly three decades of effort, the United States can take considerable pride in answering the mandates of the 1972 Clean Water Act. At great expense, much of the municipal and industrial pollution emanating from pipes, or "point" sources, has been reduced. But the largely voluntary effort to reduce "nonpoint" pollution from farms and elsewhere has been less successful. As we enter a new millennium, progress in improving water quality seems to have leveled off and our hopes of meeting the goals of the Clean Water Act seem to be receding.

Some 3,600 waterways across the nation are listed as either impaired by nutrients or by algal blooms, which are typically caused by excess nutrients. The U.S. Environmental Protection Agency is locked in a contentious legal battle with many states over additional requirements to improve water quality.

Part of the problem is the complex nature of water pollution and the difficult challenge of controlling nonpoint pollution. Clearly another part of the problem is the question of costs. States are heavily constrained by cost considerations and reluctant to embark on another costly round of stricter point source controls.

In *Fertile Ground: Nutrient Trading's Potential to Cost-Effectively Improve Water Quality*, the Director of WRI's Economics Program, Paul Faeth, proposes a way out of this dilemma. Using case studies in three states, he develops a framework to assess the cost-effectiveness of various policies and combinations of policies to reduce phosphorus loads in specific watersheds. In all three cases, policy approaches incorporating nutrient trading programs are dramatically less expensive than conventional approaches and can achieve comparable benefits.

With this report, WRI continues over a decade of work on the environmental implications of agricultural policies and practices in the United States and abroad. In *Growing Green: Enhancing the Economic and Environmental Performance of U.S. Agriculture*, Faeth and the team he managed integrated voluminous amounts of data into an analytic framework that assessed the profitability and environmental impacts of alternative cropping systems. In *Agricultural Policy and Sustainability: Case Studies from India, China, the Philippines, and the United States*, Faeth and several co-authors found that farm policies are usually stacked against resource-conserving farming methods.

WRI maintains a continuing commitment in this area. Using the methodology developed in this report, a study is currently underway that will focus on the heavily polluted "Dead Zone" in the Gulf of Mexico. WRI has also developed a prototype nutrient trading web site to help support pilot nutrient trading programs across the country.

ANTHONY JANETOS
SENIOR VICE PRESIDENT FOR PROGRAM
WORLD RESOURCES INSTITUTE

1

INTRODUCTION / OVERVIEW

The Clean Water Act of 1972 has largely succeeded in reducing pollution from industrial, municipal, and other "point" sources. But the job of achieving clean water in the United States is only about half done. Roughly 40 percent of the nation's surface waters remain at least partially impaired, and surveys suggest there has been little improvement recently.

Further progress now depends on developing low-cost, innovative approaches that can effectively cut pollution from "nonpoint" sources such as agriculture and urban runoff. But work still needs to be done to test new approaches and to determine how they can be implemented to offer cost-effective and quantifiable improvements in water quality.

Excessive nutrient loading is the single largest cause of water quality impairment in the United States.

This study amounts to a search for such new, cost-effective approaches. It is based on two premises. First, excessive nutrient loading is the single largest cause of water quality impairment in the United States; and second, attacking the worst pollution problem should yield significant benefits.

Using case studies of watersheds in three states—Michigan, Wisconsin, and Minnesota—this study develops a framework to explore the cost-effectiveness and environmental performance of various strategies to reduce phosphorus loads in specific watersheds. It describes the impact of point and nonpoint sources on phosphorus loads, the technology options for reducing those loads, and the costs of implementation using various technologies and policies. Further, it compares several policy approaches with an estimate of the least-cost outcome, giving policymakers an opportunity to compare costs relative to a theoretical minimum.

In all three cases, the analysis finds that policy approaches utilizing nutrient trading are dramatically less expensive than conventional point-source performance requirements. In the Michigan case, for example, one variation utilizing trading is estimated to cost about $2.90 per pound of phosphorus removed, compared to almost $24 per pound for conventional point-source requirements.

Based on the analysis, we recommend that

- The Environmental Protection Agency (EPA) and state environmental agencies allow trading among industrial and municipal point sources whenever new nutrient reduction requirements are put in place.

- EPA and the U.S. Department of Agriculture provide much greater regulatory and financial support for state efforts to develop point-nonpoint source pilot trading programs and standards.

- Concentrated animal feeding operations be treated the same as industrial and municipal point sources, which means that they should face involuntary permitting processes and tight enforcement.

- Sufficient money from farm income subsidy programs be shifted to conservation programs so that environmental problems caused by agriculture can be remediated using market-based approaches.

- Monitoring of water quality be vastly increased.

- Federal and state agencies explore opportunities for trading in large watersheds.

IS TRADING THE ANSWER?

Experience to date is limited, but it appears that trading can be part of the answer to achieve better water quality. Trading has been successfully used to obtain cost-effective reductions in other areas of environmental concern, including lead, sulfur dioxide, and other air emissions. In addition, trading is the leading option proposed to address the build-up of greenhouse gas emissions that could cause climate change.

When tighter standards are put in place, trading increases flexibility and reduces costs by allowing dischargers with new obligations the option of adapting their own facilities or financing comparable reductions by others. Trading makes it profitable for sources with low treatment costs to reduce their own effluents beyond legal requirements, generate a credit, and sell these credits to dischargers with higher treatment costs. This flexibility produces a less expensive outcome overall while achieving—and even going beyond—the mandated environmental target.

When tighter standards are put in place, trading increases flexibility and reduces costs. This flexibility produces a less expensive outcome overall while achieving—and even going beyond—the mandated environmental target.

This analysis examines trading as an option to achieve phosphorus reductions in three watersheds in the upper Midwest: the Saginaw Bay watershed in Michigan, the Rock River watershed in Wisconsin, and the Minnesota River Valley. In each of these cases, variations of nutrient trading are compared with traditional regulatory and subsidy approaches. The results, presented in Chapter 5, show great potential for cost savings by applying market-based strategies to water quality regulations *and* agricultural conservation subsidy programs. In Michigan and Wisconsin, pilot trading programs are being pursued, and in Minnesota, *ad hoc* trades have occurred. This report describes some of the experiences to date with these efforts and others.

While trading is the economically preferred approach for nutrient management problems, there are a number of significant concerns that must be addressed to ensure that trading is an environmentally effective and equitable solution. Some of the measures that have been suggested also affect the cost-effectiveness of the outcome, and these tradeoffs are important to recognize. A key issue in nutrient trading is equivalence. The biggest economic and environmental opportunity appears to be trading between "point" and "nonpoint" sources. But point discharges are relatively easy to quantify and monitor, while nonpoint discharges are episodic and cannot be directly measured, only estimated. However, nonpoint source reductions can also have the additional advantage of decreasing sediment loads. The apparent benefits and risks of nutrient trading as a strategy to cost-effectively support the goals of the Clean Water Act point to the need for a cautious but also encouraging approach. However, means are available to increase the certainty of outcomes in ways that still produce major savings. This and other issues of program implementation are discussed in Chapter 6.

We start by examining some of the major trends and policy issues in Chapter 2 and then explain how trading could address some of the problems in Chapter 3. In Chapter 4 we describe the three case study watersheds. This section highlights the water quality problems each watershed faces, as well as the analytical methods we used to produce our results.

The intent of the analysis is to develop and implement a policy tool to explore the cost-effectiveness and environmental performance of various strategies to improve water quality in specific watersheds.

The intent of the analysis is to develop and implement a policy tool to explore the cost-effectiveness and environmental performance of various strategies to improve water quality in specific watersheds. In doing this, we hope to illuminate the economic issues, to facilitate the development of successful pilot nutrient trading programs, and to better understand the opportunities and barriers to improving water quality under each of the policy approaches considered.

BOX 1 | WHAT TRADING IS AND ISN'T

- Trading is an adjunct to regulation, not a substitute for it.
- Trading is used to improve water quality, not degrade it.
- Trading is used between sources in the same area of impact, not outside it.
- Trading is used to reduce net loads, not increase them.

HOW DOES NUTRIENT TRADING WORK? AN ILLUSTRATION

Assume there is a river somewhere in the United States and the state environmental agency determines that the river is "impaired." There is too much phosphorus getting into the water, causing algal blooms that die and leave too little oxygen in the water for the fish to survive. The state decides to impose a limit of 1 part per million (ppm) of phosphorus in the waste stream of municipal sewage and industrial waste treatment plants. The state agrees to let these dischargers have the option of trading to meet the regulatory requirement. There are three towns along the river and a number of factories. The area also has many farms that contribute to the problem, but these aren't regulated.

The first town's sewage treatment plant is old and due for an upgrade, so that town decides to rebuild its plant and add the necessary technology to meet the standard. When bundled into the upgrade, the addition of phosphorus control is cheap, so the town's costs to meet the regulation are low.

The second town rebuilt its plant five years ago, when the state didn't require phosphorus treatment. Upgrading a relatively new plant would be expensive now, and the town wants to spend its money on a new school, not more sewage treatment. However, there is a factory nearby that also has to meet the new standard. The factory offers to reduce its phosphorus treatment to 0.5 ppm, low enough to provide the necessary credits to meet the town's requirement, if the town will split the cost of the factory's more expensive upgrade. Splitting the factory's cost is much lower than the cost of upgrading the town's facility, so the town agrees to the trade. The town and the factory both save money and when considered together, the new requirement is met for both. Since both the factory and wastewater treatment plant discharge to the river from a specific point, i.e., a pipe, this is a "point-point" trade.

The third town also upgraded its plant recently. A member of the town council gets a group of farmers to agree to put assorted conservation practices in place if the town will pay the expenses. These conservation practices not only reduce the amount of phosphorus getting into the river, but also sediment and pesticides. The state gives its approval, but the agricultural loads are variable since they only occur when it rains, so the state makes the town buy three times as much as the load reduction requirement for the plant. This "trading ratio" will provide a margin of safety for the river. Since the load of the plant is 10,000 pounds of phosphorus a year, the town must buy 30,000 pounds of reductions from the farmers or other sources that don't come from a specific point like a pipe. The cost of these "nonpoint" trades is much less than upgrading the plant, even with the extra 20,000 pounds. Rather than get all of its credits from farmers, the town council also decides to give some money to a local conservation group to restore a nearby wetland. The conservation group buys up several farms on low-lying areas of the river and plants the land with water-loving plants that grow in the region and provide habitat for waterfowl and other species. As the wetland matures, the plants trap phosphorus-carrying sediments from the water, reducing the load and improving the quality of the river for people and wildlife. The town takes credit for the reductions provided by the wetlands and applies one third of them against the load reduction mandated by the state.

2

WATER QUALITY TRENDS AND ISSUES

The principal source of information about the nation's water quality is the Environmental Protection Agency's *National Water Quality Inventory*. This is a compilation of information submitted by states, Indian tribes, territories, interstate water commissions, and the District of Columbia.

For the 1996 report, states and other jurisdictions surveyed 19 percent of the nation's river miles; 40 percent of lakes, ponds, and reservoirs; 72 percent of estuaries; 6 percent of ocean shoreline waters; and 94 percent of the Great Lakes shoreline. The survey found that:

- For rivers, 56 percent were rated good (fully supporting all uses), 8 percent were rated good but threatened for one or more uses, 36 percent were rated impaired for one or more uses, and in less than 1 percent uses were not attainable.

- For lakes, ponds, and reservoirs, 51 percent were rated good, 10 percent were good but threatened, 36 percent were impaired, and in less than 1 percent uses were not attainable.

- For estuaries, 58 percent were rated good, 4 percent were good but threatened, 38 percent were impaired, and in less than 1 percent uses were not attainable.

- For ocean shoreline waters, 79 percent were rated good, 9 percent good but threatened, and 13 percent were impaired.

- For the Great Lakes, 2 percent were rated good, 1 percent good but threatened, 97 percent impaired, and in less than 1 percent uses were not attainable.

These results are generally similar to the 1994 inventory. In that survey, surface waters only partially supporting or not supporting their designated uses included 36 percent of surveyed rivers and streams, 37 percent of lakes, ponds, and reservoirs, and 36 percent of estuaries.

Figure 1 shows the results from the EPA's surface water quality assessments undertaken between 1988 and 1996. For rivers, lakes and estuaries, the proportion of the waters surveyed that fully support their designated uses has declined, while those waters not supporting or partially supporting has increased. Although there are notable problems with this dataset,[1] the trend shown here nevertheless suggests that the United States is failing to restore the quality of the nation's water.

CAUSES OF IMPAIRMENT

The principal cause of surface water impairment is agriculture, which contributes to problems in

FIGURE 1. **PROPORTION OF SURFACE WATERS THAT SUPPORT DESIGNATED USES**

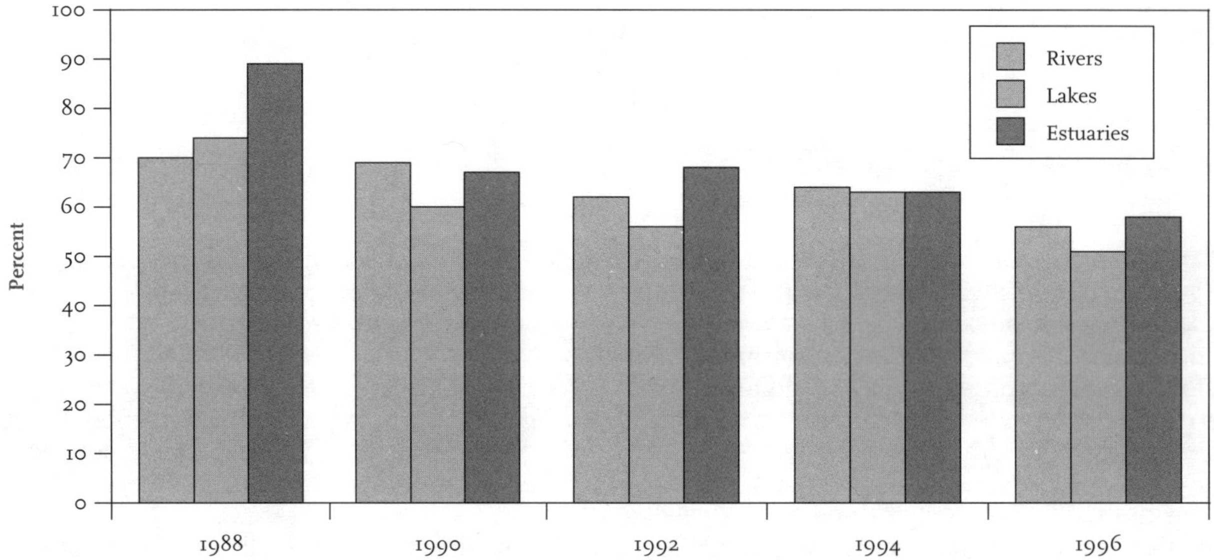

70 percent of the impaired rivers, 49 percent of the impaired lakes, and 27 percent of the impaired estuaries. *(See Table 1.)* Industrial and municipal discharges are also significant causes of water quality problems; in the nation's estuaries, they are the leading contributors to impairment. The major pollutants from these agricultural sources are nutrients, silt and oxygen-depleting substances (primarily organic wastes), metals, and bacteria. Metals have been identified as one of the leading sources of lake impairment, because of the widespread detection of mercury in the tissues of fish (USEPA, 1998).

Excessive inputs of nutrients impair water through a process called eutrophication, which causes a wide variety of problems (Smith, 1998). These include

- Increased biomass of phytoplankton;
- Shifts in phytoplankton to bloom-forming species that may be toxic or inedible;
- Decreases in water transparency;
- Taste, odor, and water treatment problems;
- Oxygen depletion;
- Increased incidence of fish kills;
- Loss of desirable fish species;
- Reductions in harvestable fish and shellfish; and
- Decreases in aesthetic value of the water body.

Nonpoint sources, particularly croplands, are by far the largest source of nutrients in Ameri-

TABLE 1. **LEADING SOURCES AND POLLUTANTS OF IMPAIRED WATERWAYS AFFECTING TWO OR MORE CATEGORIES OF SURFACE WATERS**

	Percentage		
	Rivers	**Lakes**	**Estuaries**
Sources of Impairment			
Agriculture	70	49	27
Industrial point sources/discharges	14	–	56
Municipal point sources	9	18	44
Urban runoff/Storm sewers	13	21	20
Land disposal of wastes	–	11	19
Hydromodification	14	14	–
Pollutants/Stressors			
Nutrients	40	51	57
Siltation	51	25	–
Oxygen-depleting substances	29	21	33
Bacteria	32	–	42
Metals	16	51	–
Habitat alterations	19	–	14

Source: USEPA, 1998. National Water Quality Inventory.

Nonpoint sources, particularly croplands, are by far the largest source of nutrients in American waterways.

can waterways. (*See Table 2.*) One study found that nonpoint sources account for 82 percent of total nitrogen discharges and 84 percent of total phosphorus discharges. Croplands alone account for 39 percent of all nitrogen loads and 30 percent of all phosphorus loads. In contrast, point sources account for 18 to 19 percent of nitrogen and phosphorus discharges (Carpenter et al., 1998).

Farmers have dramatically increased nitrogen fertilizer use since the mid-1960s. (*See Figure 2.*) Nitrogen use has increased by about 150 percent, from 4.6 million tons in 1965 to about 12 million tons in the mid-1990s. Phosphorus applications in the form of P_2O_5 were about 25 percent higher in 1995 than in 1965; use peaked in 1979 at 5.6 million tons (USDA, 1997a).

Plants do not use a large percentage of the nutrients applied. A National Research Council analysis of nutrient use in the United States found that up to 40 percent of nitrogen and about 60 percent of phosphorus is not taken up by crops (National Research Council, 1993). In the case of nitrogen, which is soluble, most of the residual is readily available and contaminates

TABLE 2. **NITROGEN AND PHOSPHORUS DISCHARGES TO SURFACE WATERS FROM POINT AND NONPOINT SOURCES IN THE UNITED STATES** (IN THOUSANDS OF METRIC TONS PER YEAR)

Source	Nitrogen	Phosphorus
Nonpoint sources		
Croplands	3,204	615
Pastures	292	95
Rangelands	778	242
Forests	1,035	495
Other rural lands	659	170
Other nonpoint sources	695	68
Total nonpoint discharges	**6,663**	**1,658**
Total point sources	**1,495**	**330**
Total discharge (point + nonpoint)	8,158	2,015
Nonpoint as a percentage of total	**82**	**84**

Source: Carpenter et al., 1998.

FIGURE 2. **FERTILIZER USE 1965–1995**

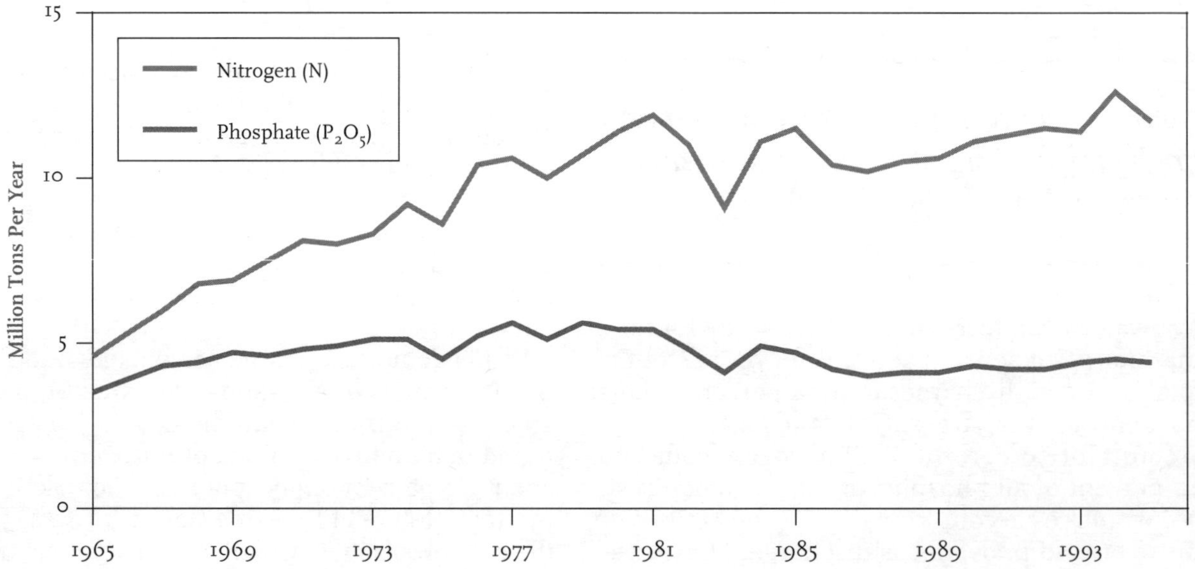

the environment through runoff, leaching, and volatilization to the atmosphere. Phosphorus binds to the soil and reaches surface waters through soil erosion processes. So much phosphorus has been used in the past that in many areas extension agents recommend that farmers stop applying it.

Animal manure is another key source of nutrient use on croplands and pastures. In the last 30 years, and particularly in the last 10, livestock populations have shifted away from cattle and toward poultry, resulting in greater nutrient concentration in wastes. Poultry litter has a much higher concentration of nutrients per unit than cattle manure, roughly 150 percent more nitrogen and 200 percent more phosphorus. The total population of dairy cows and other cattle is down by 12 percent since 1972, while the population of poultry layers and broilers is up by 134 percent. The number of turkeys has increased by 180 percent.[2] The volume of animal wastes has not changed between 1972 and 1996, but the nutrient levels in waste have gone up significantly. Using livestock numbers and standard waste and nutrient coefficients, we estimate that the total amount of nitrogen in domestic livestock wastes is up by 17 percent and phosphorus by 18 percent. [3]

The total availability of nitrogen and phosphorus from fertilizer and animal wastes in 1995 was 21.4 million tons and 7.4 million tons, respectively. This represents no change in the case of phosphorus, but a 30-percent increase in the availability of nitrogen for cropland applications from these sources since 1972.

A contributing factor in water pollution from animal wastes is that manure is typically applied well beyond crop needs. About 90 percent of the manure from animal operations does not leave the farm where it is produced (Bosch and Napit, 1992); because of high transportation costs, it is usually applied only to the land on the farm. It typically takes an area of cropland about 1,000 times greater than the feedlot to distribute manure at rates comparable to crop requirements (NRC 1993). If manure is applied excessively, phosphorus builds up in the soil, and without proper erosion control will run off into surface waters. Excess nitrogen not only runs off into waterways, but can also leach and contaminate groundwater or escape to the atmosphere as nitrous oxide (N_2O), a gas that contributes to global warming.

Another trend contributing to excessive nutrient flow into the environment is lagging sewage treatment. Because of population growth, there are now more people who are not served by some form of sewage treatment than in 1980. The percentage of people served from 1980 to 1996 increased from 70 to 72 percent (USEPA, 1996 and 1980 Needs Surveys). However, population increased over that same time by 38 million people, or 17 percent (U.S. Department of Commerce, 1998). According to our calculations, about 75 million people—up by 5 million since 1980—do not currently have access to sewage treatment.

The Clean Water Act set a national goal of achieving zero discharge by 1985. But most of the evidence suggests that we are moving further from this goal, rather than closer towards it.

WATER QUALITY POLICY ISSUES

The Clean Water Act (CWA) set a national goal of achieving zero discharge by 1985. But most of the evidence suggests that we are moving further from this goal, rather than closer towards it.

Why are we failing to achieve the goals of the Act? One reason is that the problem has grown. With more people, a bigger economy, and more demand on natural resources as receptors of pollution and providers of water, recreation and other services, the challenge has grown larger.

A second major reason concerns policy formulation. In the water quality area, policy is fractured, expensive, and inconsistent. Both the Environmental Protection Agency and the Department of Agriculture (USDA) address water quality problems at the federal level. As a broad generalization, given the relative levels of effort, the focus of these two arms of the government can be characterized as regulation to achieve point source control and subsidies for voluntary nonpoint source abatement, though both use regulation, grants, and subsidies to some extent.

When the Clean Water Act was first passed, several dramatic events—such as the burning of the Cuyahoga River in Ohio—prompted a national focus on controlling industrial and municipal sources. Under the National Pollution Discharge Elimination System (NPDES), state environmental agencies with USEPA oversight grant permits to point source dischargers such as municipal and industrial wastewater treatment plants. These permits control the discharge of pollutants into the nation's waters. The permits specify control levels that must be achieved. The agencies also monitor sources for compliance and can take a variety of actions to force compliance when permitted discharge levels are violated.

A great deal of money has been spent in the United States on municipal and industrial waste treatment and agricultural conservation practices. For example, in the original 1972 Act the federal government agreed to pay up to 75 percent (later reduced to 55 percent) of the construction and design cost for municipal treatment plants. Between 1974 and 1994, about $96 billion was spent through the federal construction grants program for new municipal construction and upgrades for point source control. Local governments have added another $117 billion (AMSA/WEF, 1999).

Over the next 20 years, EPA estimates that almost $140 billion in capital costs will be needed for municipal treatment works and related needs . . . another $190 billion will be needed by local governments to replace aging facilities and collection systems.

Over the next 20 years, EPA estimates that almost $140 billion in capital costs will be needed for municipal treatment works and related needs. The Association of Metropolitan Sewerage Agencies (AMSA) and the Water Environment Federation say that another $190 billion will be needed by local governments to replace aging facilities and collection systems, not including operation and maintenance costs (AMSA/WEF, 1999).

The approach taken for nonpoint sources stands in stark contrast to that for point sources. Abatement programs for nonpoint source pollution occur mainly through subsidy programs provided by the USDA and EPA. The lion's share of the funds come through agricultural legislation to farmers for land retirement and cost-share programs, primarily for erosion control. In recent years, the USDA has spent around $3.5 billion per year on conservation programs, extension, administration, and research. EPA spends about another $800 million on its voluntary nonpoint programs

(USDA, 1997a). Approximately half of the USDA's money goes to the Conservation Reserve Program (CRP), which was conceived primarily as a means to keep land out of production to support crop prices. Environmental benefits of the CRP were initially focused on reducing soil erosion; in this respect, the program has been effective. Since 1982, soil erosion in the United States has been cut by 40 percent (USDA, 1994a). Adjustments in the program have attempted to make it more responsive to water quality needs, but about 60 percent of the acreage in the program is concentrated in the Northern Plains, Southern Plains, and Mountain States, where water quality benefits from the program are relatively small (USDA, 1997a).

There is some measure of involuntary regulation of nonpoint sources provided in the CWA, particularly for farms larger than 1,000 animal units.[4] However, even for such large operating units, regulation is not automatic in most states. The state first has to make a determination that the source is discharging to a water body and contributing to a water quality problem; only then will the farm be required to apply for a permit. Of 1.1 million farms in the United States, just 10,000, or 1 percent, have permits under the CWA (Adler, et al., 1993).

Wastes from animal operations are not controlled with anywhere near the same rigor as human wastes, even though, in the United States, waste from livestock is about 130 times greater than that from humans.

As a result, wastes from animal operations are not controlled with anywhere near the same rigor as human wastes, even though, in the United States, waste from livestock is about 130 times greater than that from humans. Taken together, the 1,600 dairies of California's Central Valley produce waste equivalent to 21 million people. The 600 million chickens of the Delmarva Peninsula on the eastern side of the Chesapeake Bay generate as much nitrogen as a city of 500,000 (U.S. Senate, 1997).

Yet, there is a wide variance across the states regarding nonpoint source regulations and

BOX 2 | THE CLEAN WATER ACT

Federal water pollution control law attempts to protect water quality by controlling the discharge of pollutants. Congress passed the Clean Water Act (CWA) in 1972 with the basic objective "to restore and maintain the chemical, physical and biological integrity of the Nation's waters." The CWA sought to have "fishable and swimmable waters" by 1983, "zero discharge" of pollutants by 1985, and to eliminate the release of "toxics in toxic amounts." The law made the EPA responsible for setting national standards for the discharge of effluents on an industry-by-industry basis, considering both the capabilities of pollution control technologies and the costs of implementation. This legislation makes it illegal to discharge pollutants to surface waters without a permit. The legislation specifically focused on point sources. The law was extensively amended in 1977 and 1987 to expand EPA's powers and to address nonpoint pollution through voluntary programs.

Adapted from: Adler et al., 1993, and Arbuckle et al., 1993.

their implementation (Environmental Law Institute, 1997). As of 1996, only 16 states required nutrient management plans, and these were usually in areas suffering from groundwater contamination. Eighteen states required practices for controlling soil erosion, but only when a complaint is filed by a citizen or government agency. Implementation in response to the complaint is only required if cost-share funds are available from the government (USDA, 1997a).

It is clear that the nation's existing water quality policy framework is inadequate. Not only is the nation failing to meet the goals set out by Congress, but forward progress seems to be slow and difficult. In many cases, environmental groups have resorted to litigation in an attempt to make state and federal agencies enforce the basic requirements of the CWA.

States are charged with identifying waters within their jurisdiction that do not meet designated uses and developing Total Maximum Daily Loads (TMDLs) to address the problems. TMDLs are plans to establish and allocate loading targets. This typically means additional requirements for point sources and more subsidies for nonpoint sources. Where states do not accomplish these tasks, EPA must do it for them. There are currently 12 states where EPA is under court order to establish TMDLs if the states do not, another 15 states with pending litigation, and 4 states where notices of intent to sue have been filed (USEPA, 1999).

The list of impaired waterways is extensive. Section 303(d) of the CWA requires states to identify waters that are not fishable or swimmable and to develop TMDLs for these waters. Nationwide, there are 3,456 waterways listed as impaired by nutrients and another 141 impaired by algal blooms, typically caused by excess nutrients. (*See Table 3.*)

Nationwide, there are 3,456 waterways listed as impaired by nutrients and another 141 impaired by algal blooms, typically caused by excess nutrients.

TABLE 3. **THE TOP 20 STATES BY NUMBER OF WATERWAYS IMPAIRED BY NUTRIENTS**

Illinois	634
Florida	539
Mississippi	469
Oklahoma	218
Pennsylvania	217
Ohio	204
Montana	156
Maryland	145
Delaware	138
Massachusetts	135
North Dakota	98
Tennessee	60
California	58
Kentucky	52
Wisconsin	46
Connecticut	36
Rhode Island	35
Michigan	35
Iowa	25
Nebraska	24
Subtotal	**3,324**
Total 50 States	**3,456**

Source: USEPA, 1999.

3

WHAT ABOUT TRADING?

Many studies on the economics of environmental regulation emphasize the importance of allowing flexibility in addressing environmental problems.

Trading, as an adjunct to regulation, can help provide the flexibility to keep compliance costs down while meeting or exceeding water quality goals.

Trading, as an adjunct to regulation, can help provide the flexibility to keep compliance costs down while meeting or exceeding water quality goals. For example, discharge permits, a common regulatory approach for reducing point source pollution, specify the amount and type of allowable discharge to the nation's waters. In the case of nutrients, permits are used to limit discharges that were previously uncontrolled or tighten prior limits, causing permit holders to incur significant capital and operating costs to comply. For several reasons, including size, control technology, and characteristics of the incoming waste stream, dischargers can face vastly different costs of compliance, but have little flexibility to seek cost-effective solutions without programs such as trading.

There are two fundamental types of trading programs for any type of discharge: open and closed. Closed trading programs are far more common and are an extension of traditional regulation in that they begin with a mandate to reduce discharges. Often called "cap and trade," these systems include a mandatory "cap" on emissions or discharges and individual allowances to sources within a defined trading area. A cap is established by the regulatory agency to achieve or maintain ambient air, water, or other environmental quality standards. A TMDL—Total Maximum Daily Load—is the process under the Clean Water Act that establishes and allocates the cap within a watershed or impaired area. Caps may be fixed or declining depending on the attainment or nonattainment status. A cap can be increased without undermining the integrity of the water quality objectives if sources not controlled by the cap opt in and reduce their loads by at least the amount that the cap is extended. For example, if a cap were on point sources only, nonpoint sources could make load reductions that could be used to increase the point source cap. Allocations in the form of allowances are established for each discharge site within the trading area based on a percentage of overall reductions that must be achieved from a baseline of actual or allowed historical discharge levels. A trading system is said to be "fully closed" when all discharge sites are controlled under the cap and the cap is equal to the total permissible load for the watershed (Stephenson and Shabman, 1996).

Closed systems have been developed and implemented in areas where ambient environmental standards are not being met to provide a more cost-effective means of achieving the reductions necessary to attain these standards.

The Acid Rain Trading Program under the federal Clean Air Act Amendments of 1990 is one example of a closed trading system. The program, which set a cap of 8.95 million tons of SO_2 per year, is enforced through a system of tradable emission allowances. From 1995 to 1997 the program exceeded expectations, with firms over-achieving the reduction target at less than half the cost. Industry and EPA initially estimated that abatement costs would be in the range of $750-$1,000 per ton, but allowance prices reached a low of $66 per ton in mid-1995 and have generally remained below $200 per ton since the system's inception. Over the long run, it is estimated that allowance trading could result in savings of $700 to $800 million per year compared to a command-and-control approach with a uniform emission standard (Anderson, 1999).

The states of California and Illinois have also implemented closed air emission trading programs designed to attain the national ambient air quality standard for ozone.

Open systems are usually voluntary. Where ambient standards are being met, they are used to maintain or improve environmental quality while allowing for economic growth and development. A number of states are looking into the development of open air and water quality trading programs. Michigan's air emission trading program took effect on March 16, 1996, and the states of Florida and Texas are also working to implement similar programs.

Open trading systems rely on existing regulations to establish a baseline; reductions from the baseline generate a reduction credit. Credits can be banked, traded, or used to comply with discharge limits established by an applicable regulation. Open systems offer greater operational flexibility and improvements in environmental quality without establishing a mandatory cap and allowance for each discharge site. Reductions must be real, surplus, and enforceable to be used or traded under an open system.

In a few instances, regulatory agencies have allowed trading between polluters. In these cases, those sources with high remediation costs have been allowed to purchase unused emissions allowances provided under another facility's permit. The most important determinants of success have been cost and sufficiency of trading options, limited regulatory restriction on trading, low transaction costs, certainty in establishing pollution levels, and anticipated improvement in environmental quality (Hahn and Hester, 1988).

Regulators have allowed trading in a small number of cases for water quality management (Bacon, 1992; Hahn and Hester, 1988; USEPA, 1996). While effluent trading programs have been very successful for lead, air emissions, and sulfur dioxide trading, the results in the water quality area are not impressive.[5] In the first instance, the state of Wisconsin gave sources such as wastewater treatment plants, and pulp and paper mills the right to meet state water quality standards through the trading of effluent rights. The state initiated biological oxygen demand (BOD) trading on a 35-mile stretch of the heavily industrialized Fox River and on 500 miles of the Wisconsin River. It allowed trades between the paper mills and municipal wastewater treatment plants. However, regulators would only allow trades if permitted discharge levels could not be met through the adoption of technological standards. Cost reduction was not considered a legitimate reason for trading. Since 1981, only one trade has occurred. Several factors are thought to have contributed to the limited activity, including the severe restrictions imposed by the state on the ability of sources to trade, the vulnerability of the program to legal challenge, and the fact that the dischargers developed a variety of compliance alternatives that had not been foreseen when the regulations were drafted (Anderson, 1999).

The Environmental Protection Agency is currently considering extending point-point trading for copper dischargers in San Francisco Bay and for dischargers of nitrogen and suspended solids in Tampa Bay.

POINT-NONPOINT TRADING

In the point-nonpoint trading area, programs are currently in place at the Dillon and Cherry Creek Reservoirs in the Denver, Colorado, area; in North Carolina's Tar-Pamlico Basin; and are being considered in many other places.[6]

Dillon Reservoir provides about half of the city of Denver's water supply. Four municipal wastewater plants discharge into the reservoir. Studies in the 1980s indicated that phosphorus discharges would have to be reduced to maintain the reservoir's water quality and accommodate future growth, and that reductions would have to include nonpoint sources such as runoff from lawns and streets, and seepage from septic tanks. Using 1982 phosphorus discharges as the baseline, the plan established a cap on total phosphorus loadings, set load limits for the four treatment plants, and allowed trading of phosphorus loadings with nonpoint sources (Anderson, 1999).

In this case, nonpoint source control was allowed as a means to supplement point source management, provide for regional growth, and improve water quality in the reservoir. Point sources can increase discharges above approved limits only if they reduce loads from nonpoint sources in the watershed. In order to deal with the uncertainty of nonpoint source reductions, only half of the expected reduction from nonpoint sources can be applied against point source loads. Since 1984, point sources in the area have increased their efficiency and reduced their phosphorus loads significantly. Point sources have generally been able to keep their loads below regulatory limits, so only three trades have taken place. Future trades are expected as development in the area proceeds (Bacon, 1992; Michigan DEQ, 1998a). A similar program is in place at Cherry Creek Reservoir—also a source of water for the Denver region—but no trades have taken place to date because phosphorus loadings at municipal treatment facilities are still below the limits set by the Colorado Water Basin Commission.

In 1989, the North Carolina Environmental Management Commission designated the Tar-Pamlico Basin as "nutrient-sensitive" after studies found that algae blooms and low dissolved oxygen posed a threat to the basin's fishery resources. The formal designation required the state Division of Environmental Management (DEM) to identify the nutrient sources, set nutrient limitation objections, and develop a nutrient control plan. The DEM analysis showed that most of the basin's nutrient loadings (primarily nitrogen, but also phosphorus) came from agricultural runoff and other nonpoint sources. Other sources include municipal wastewater treatment plants and industrial and mining operations (Anderson, 1999; Michigan DEQ, 1998a).

In the Tar-Pamlico, an association of point source dischargers is allowed to trade with one another under a cap. If the cap cannot be met, members of the association can pay into a fund to support a government-managed program that encourages the adoption of best management practices by farmers in the watershed. Tar-Pamlico is a hybrid of a trading program and an effluent tax, since the credits are purchased at a fixed price and there is no direct connection between the credits needed by the point sources and the credits generated by the nonpoint source program. In the first phase of the program, association members improved the efficiency of their facilities and successfully met defined standards. Allowable discharges are being gradually reduced during the second phase, which began in 1995 (USEPA, 1992; Michigan DEQ, 1998a).

A number of other trading programs are in existence or being developed in Colorado, Connecticut, Michigan, Minnesota, and Wisconsin. However, in the existing cases, only limited markets for nutrient trading have emerged because the programs are not fully functional or allowable discharge limits have not been reached. Some programs have actually increased administrative and transaction costs, making trading difficult.

The opportunities for nutrient trading appear to be enormous. A conservative estimate suggests that point-nonpoint trading could be applicable in as many as 900 watersheds in the United States.

THE POTENTIAL FOR POLLUTION TRADING

On a strictly physical basis, the opportunities for nutrient trading appear to be enormous. A conservative estimate suggests that point-nonpoint trading could be applicable in as many as 900 watersheds in the United States (USEPA, 1992a). Nonpoint sources, in particular agriculture, are major contributors to water pollution and remain largely unregulated, creating a vast opportunity for significant reductions in polluted runoff. The U.S. National Research Council estimated national nitrogen "residuals"—that is, nitrogen not taken up by plant material—on cropland at 6.6 to 9.1 million metric tons and phosphorus residuals at 2.9 million metric tons, much of which represents excess use. Other researchers have shown that as much as three quarters of excess nutrient use can be avoided through fairly simple and inexpensive techniques such as soil testing and fertilizer banding (NRC, 1993).

From an economic standpoint, the opportunities for cost savings from point-nonpoint source trading are large. One estimate suggested the cost of point-source reduction could be 65 times higher than nonpoint source reduction (Bacon, 1992). EPA estimates that if background pollution from agriculture were reduced, the need for tertiary (advanced) water treatment could be avoided—providing a net saving of $15 billion in capital costs for tertiary treatment (USEPA, 1992).

Nutrient trading between point sources makes sense as well. Wastewater treatment, like many processes, is characterized by significant economies of scale; larger facilities can process wastewater at much lower per-unit cost than small facilities. Particularly in cases involving a large number of small treatment works, trades involving a few large operations that overcomply and sell the credits to the smaller ones can dramatically lower overall compliance costs while achieving the same environmental objective.

An analysis of President Clinton's 1994 Clean Water Initiative (CWI) suggested that nutrient

An analysis of President Clinton's 1994 Clean Water Initiative suggested that nutrient trading could lower the cost of implementing the initiative by $658 million to $7.5 billion.

trading could lower the cost of implementing the initiative by $658 million to $7.5 billion, based on total incremental costs estimated to range from $5 billion to $9.6 billion. The bulk of these savings (75 to 92 percent) would be realized from point-nonpoint trading, while the remaining savings were expected to come from point-point trading (USEPA, 1994).

Two other possible options for nutrient trading include pretreatment trading, where a municipal plant pays a facility that is discharging into its system to reduce its contribution rather than upgrading the system itself. Nonpoint-nonpoint trading, another option, is possible in theory, but practical examples do not exist.

In short, nutrient trading may represent a vast untapped reservoir to resolve the nation's water quality problems, but significant implementation and performance issues must be tackled.

4

NUTRIENT TRADING OPPORTUNITIES IN THREE WATERSHEDS OF THE UPPER MIDWEST

Nutrient trading is widely thought to reduce the costs of improving water quality, but there is little economic analysis comparing trading with other policy approaches. Of the work done to date, most studies simply compare average costs of various forms of nutrient reduction for point and nonpoint sources.

These three cases were chosen because each area has significant water quality problems and the responsible state agency was interested in exploring nutrient trading as a policy alternative.

In the course of this study, we developed a comprehensive analytical framework to compare the economic and environmental performance of alternative policy strategies to reduce nutrient loads. In particular, we explored opportunities to reduce phosphorus loads in three watersheds of the Upper Midwest: the Saginaw Bay in Michigan, the Rock River in Wisconsin, and the Minnesota River Valley. (*See Figure 3.*)

The intent of the analysis was to develop and implement a policy tool to explore the cost-effectiveness and environmental performance of various strategies to improve water quality in specific watersheds. In doing this, we hoped to illuminate the economic issues, to facilitate the development of successful pilot nutrient trading programs, and to better understand the opportunities and barriers to improving water quality under each of the policy approaches considered.

THE CASES: NUTRIENT TRADING IN MINNESOTA, MICHIGAN, AND WISCONSIN

These cases were chosen because each area has significant water quality problems and the responsible state agency was interested in exploring nutrient trading as a policy alternative. The study is a joint effort of the World Resources Institute; the Michigan, Wisconsin, and Minnesota state environmental protection agencies; and various stakeholders in each watershed, including point source operators, farmers, environmental groups, and academics. In each of these three states, there has been significant movement toward the use of trading.

Background

Notwithstanding requirements to upgrade point source discharges and the development of nonpoint source programs, nutrient pollution in the Minnesota River has been a chronic problem for years. In 1988, the Minnesota Pollution Control Agency (MPCA) established a Total Maximum Daily Load (TMDL) on the lower 25 miles of the

FIGURE 3. | CASE STUDY WATERSHEDS

Minnesota River, limiting chemical and biological oxygen demand (BOD) and ammonia nitrogen (USEPA, 1992b). Phosphorus control is now under active regulatory consideration by the MPCA. A study of point source treatment found that incremental costs for phosphorus control range from about $15 million to $39 million per year (Metcalf and Eddy et al., 1993). A preliminary study done for EPA in 1987 identified the basin as a good candidate for a trading program (Industrial Economics, 1987).

Recently this interest materialized in the form of a trade tied to a new permit. In January 1997, Minnesota Governor Arne Carlson announced a new discharge permit for the Rahr Malting Company that allows the company to build its own wastewater treatment facility on a tightly regulated stretch of the Minnesota River in exchange for reducing nonpoint source discharges upstream from the plant (MPCA, 1997a). Under the terms of the agreement, Rahr can support agricultural conservation practices including soil erosion controls, livestock exclusion from waterways, rotational grazing, critical-area set asides, and wetland treatment systems for nutrient removal. Credits that Rahr receives from these practices can be applied against the load of its new facility. Detailed papers outline how the amount of credit from various practices is to be calculated (MPCA, 1997b, 1997c,1997d).

Michigan has developed rules for a voluntary statewide "water quality" trading program and is conducting a demonstration project on the Kalamazoo River.

Michigan has developed rules for a voluntary statewide "water quality" trading program and is conducting a demonstration project on the Kalamazoo River.[7] The Kalamazoo project is intended to help the state and a group of stakeholders obtain design information for development of the statewide program (Michigan DEQ, 1998a). The Water Quality Trading Group in Michigan has drafted an extensive Water Quality Trading Rule to enable it to move forward and implement the demonstration program (Michigan DEQ, 1998b). The draft has been submitted to the Region V office of the USEPA; discussions are continuing on the merits of the draft rule.

Though Michigan ultimately chose to implement a pilot program in the Kalamazoo River watershed, Saginaw Bay was chosen as the most appropriate case for this study. Saginaw Bay has been the focus of water quality improvement efforts by the state of Michigan for more than 15 years. The state has sponsored several studies and made major efforts to reduce nutrient loads; the latest effort is the Saginaw Bay Soil Erosion and Sedimentation Control Program. Because Saginaw Bay has been the subject of prior studies, the information necessary to complete a research effort on nutrient trading was generally available and comparative analysis was possible. According to Department of Environmental Quality (DEQ) staff, the analysis presented here was a key factor in the state's determination to undertake a statewide program (Batchelor, 1999).

According to both the Wisconsin Department of Natural Resources and the U.S. Environmental Protection Agency, the Rock River is a water body that does not attain its designated uses. As such, the river is subject to a limit on the allowable load of pollutants it can receive, including phosphorus.

In response to a proposed rule to require all the point sources in the Rock River watershed to adopt phosphorus controls, a group of point source dischargers called the Rock River Partnership petitioned the secretary of the Wisconsin Department of Natural Resources to grant them an extension so that a nutrient trading program could be tried in the watershed. The partnership has assessed its membership according to their size in order to raise funds to pay for hydrological modeling and monitoring of the basin. In 1996, the governor's budget included funds to study nutrient trading in four watersheds, the Rock River among them. These pilot programs are intended to help develop a statewide framework for trading (Michigan DEQ, 1998a).

Table 4 summarizes EPA's vulnerability indicators for the 23 sub-basins found in the three watersheds in the present study. The data show the nature of the water quality problems. Just one of the sub-basins have "best" or "better" condition, while three quarters have "more serious" conditions (a five or six rating, with six

TABLE 4. SUB-BASIN VULNERABILITY INDICES BY WATERSHED

Sub-Basin Name	Overall Condition	Aquatic Species at Risk	Urban Runoff Potential	Index of Agricultural Runoff Potential	Population Change	Hydrologic Modification
Saginaw Bay Watershed						
Au Gres-Rifle	1	Low	Low	Moderate	**High**	Moderate
Kawkawlin-Pine	5	n/a	Low	Moderate	Low	Low
Pigeon-Wiscoggin	5	Moderate	Low	**High**	Low	Low
Tittabawassee	3	Low	Low	Moderate	Moderate	Moderate
Pine	5	Low	Low	**High**	Low	Moderate
Shiawassee	5	Moderate	Low	**High**	Moderate	Moderate
Flint	4	n/a	Low	**High**	Low	Moderate
Cass	5	Low	Moderate	**High**	Low	Low
Saginaw	5	Low	Low	Moderate	Low	n/a
Minnesota River Watershed						
Upper Minnesota	3	Low	Low	**High**	Low	Moderate
Pomme De Terre	5	Low	Low	**High**	Low	Moderate
Lac Qui Parle	5	n/a	Low	**High**	Low	Moderate
Hawk-Yellow Medicine	3	Moderate	Low	**High**	Low	Moderate
Chippewa	5	n/a	Low	**High**	Low	Moderate
Redwood	5	n/a	Low	**High**	Low	Low
Middle Minnesota	5	Moderate	Low	**High**	Moderate	Moderate
Cottonwood	5	Moderate	Low	**High**	Low	Low
Blue Earth	5	Moderate	Low	**High**	Low	Moderate
Watonwan	5	n/a	Low	**High**	Low	Moderate
Le Sueur	5	n/a	Low	**High**	Low	Moderate
Lower Minnesota	6	**High**	Low	**High**	**High**	Moderate
Rock River Watershed						
Upper Rock	4	Moderate	Moderate	**High**	**High**	**High**
Crawfish	5	Low	Moderate	**High**	Moderate	Moderate

Source: Environmental Protection Agency, "Surf Your Watershed" website.

Notes: The overall condition describes the health of the aquatic resources for each watershed. The score is the result of combining 15 indicators of watershed condition and vulnerability, zero being the best water quality or condition, and 6 being the most serious condition.

being the worst possible rating). Just one sub-basin is highly vulnerable to urban runoff, while 19 are highly vulnerable to agricultural runoff. A few sub-basins are vulnerable to "population change" (meaning growth) and the indicators show that many are moderately vulnerable to hydrologic modification such as dams. Not shown here, Kawkawlin-Pine is highly vulnerable to conventional loads over permitted levels. Most other basins have low vulnerability to toxic or conventional loads over permitted levels. For various reasons, only one sub-basin attains its designated uses.

Point Sources and Control Options

The three watersheds have a number of things in common and a few important differences. Three key differences are the number of point sources in each watershed, the variation in their size distribution, and their level of treatment. Figure 4 shows the number and size distribution for each watershed. In Minnesota, 211 facilities release phosphorus into the Minnesota River. They range in size from a few thousand gallons per day to 26 million gallons per day (mgd), with a total flow of 130 mgd. The two largest municipal plants in the region have flows of 26 mgd and 21 mgd. The vast majority of the plants on the river are considerably smaller; most of the flow is concentrated in a few facilities. Of the total, 199 have a flow of less than 1 mgd and 126 have a flow less than 0.1 mgd. Only the two largest plants currently have phosphorus control.

The average concentration of phosphorus in effluent from all point sources in the watershed is 1.76 parts per million[8] (ppm), according to estimates made using Minnesota Pollution Control Agency (MPCA) data. The total phosphorus

FIGURE 4. DISTRIBUTION OF POINT SOURCE DISCHARGERS IN CASE STUDY AREAS BY FLOW RATE

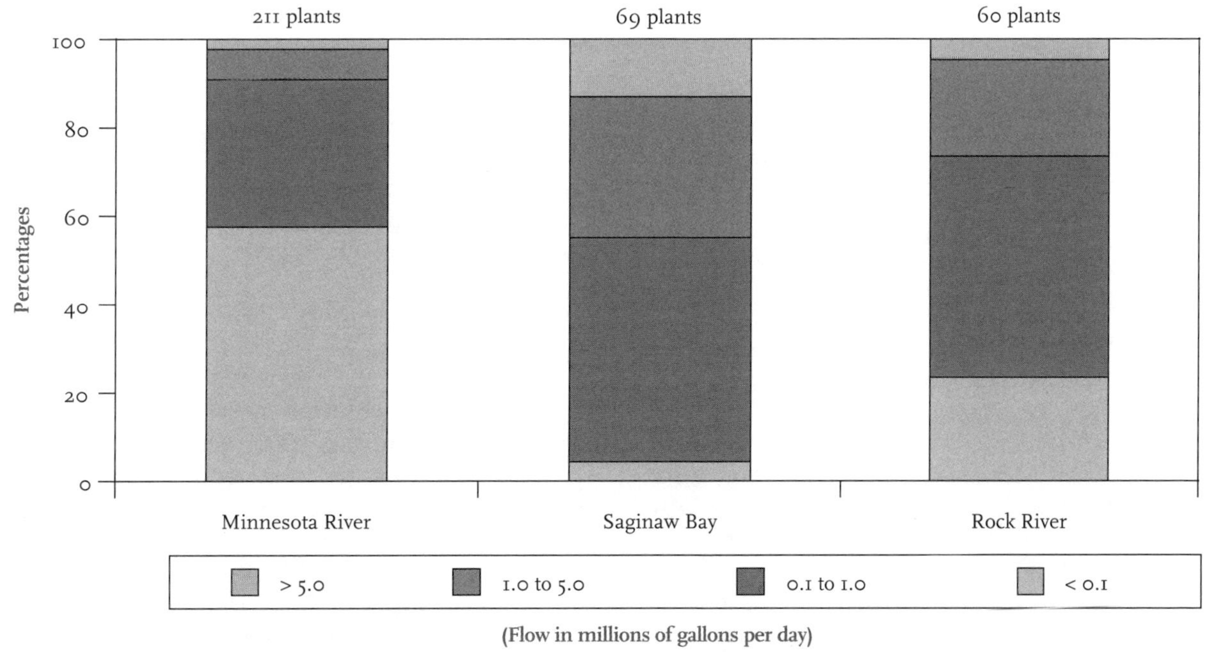

load to the watershed from these sources is about 348 tons per year.

Sixty point source dischargers release phosphorus into the Rock River in Wisconsin. They range in size from 30,000 gallons per day to 40 million gallons per day at Madison, which is by far the largest plant. The majority of the plants on the river are small, but there is a much greater proportion of large plants than in Minnesota. Of the total, 16 have a flow of more than 1 mgd and 35 have a flow less than 0.25 mgd. Only Madison's municipal wastewater treatment plant has phosphorus control. Under the proposed state regulation (NR217), plants with loads less than about 1,300 pounds per year would not have to reduce P loadings. This means that 25 of the 60 dischargers would have no obligation to comply under the proposed rule.

The average concentration of phosphorus in effluent from point sources in the watershed subject to the regulation is about 2.7 ppm, according to estimates derived from Wisconsin Department of Natural Resources (WDNR) data. The total phosphorus load to the watershed from these sources is about 401 tons per year. The point sources that are not subject to the regulation contribute an additional 14 tons per year of phosphorus, or 3 percent of the total point source load.

The Minnesota and Rock Rivers drain into the Mississippi, but Saginaw Bay is part of the Great Lakes system. For this reason, only the facilities in that watershed have had to adopt treatment for phosphorus removal. There are 69 point source dischargers with permits to release phosphorus into the Saginaw Bay Watershed. They range in size from an average of a few thousand gallons per day to 89 million gallons per day. The largest municipal wastewater treatment plants in the region are Saginaw (25 mgd) and Flint (28 mgd). There are few plants smaller than 0.1 mgd in the basin and a large proportion of relatively large facilities.

By law, the standard concentration limit for phosphorus in effluent that flows into the Great Lakes is 1 ppm, though some plants have tighter restrictions or have voluntarily achieved lower limits. The average concentration of phosphorus in effluent from all point sources in the watershed is 0.74 ppm, according to data reported to the Michigan Department of Environmental Quality (MDEQ) by dischargers. The total phosphorus load to the watershed from these sources is about 321 tons per year.

Point Source Controls and Costs

For sources that have been subject to phosphorus limits for some time, as in Michigan, the standard method of control has been chemical phosphorus removal. In this method, metal salts of aluminum and iron are added to waste streams to form compounds that precipitate phosphorus from the wastewater. For effluent limitations of 1 ppm, metal salt additions in conjunction with conventional clarifiers do an acceptable job. It is possible to reduce phosphorus levels to 0.5 ppm in the effluent by adding more metal salt, but to achieve limits of 0.2 ppm it is necessary to use secondary filtration with either chemical or biological treatment. (*See Table 5.*)

Biological phosphorus removal is a relatively new technology. There are at least six different types of biological phosphorus removal systems; all rely on biota to remove excess phosphorus from the waste stream during their growth processes. For most point sources with existing chemical treatment, biological phosphorus removal cannot be technically accommodated, so these plants are locked into higher control cost options. Because costs are lower, most new plants use biological treatment processes.

TABLE 5. NUMBER OF PLANTS IN EACH WATERSHED BY PHOSPHORUS (P) TREATMENT TYPE

	Minnesota River Valley	Saginaw Bay	Rock River
No treatment	209	21[a]	34[b]
Standard chemical	0	21	0
Maximum chemical	0	17	0
Chemical with filtration	0	10	0
Biological	2	0	1
Biological with filtration	0	0	0
Total for Watershed	211	69	60
Average concentration of P in effluent (ppm)	1.76	0.74	2.7
Total Point Source P Load	348	321	401

Notes:

a. These sources meet the discharge standard without treatment.

b. Of this number, 25 plants are too small to meet the minimum compliance requirements.

For this analysis, we characterized six categories of treatment:

1. no treatment;
2. standard chemical phosphorus removal;
3. maximum chemical phosphorus removal possible without filtration;
4. chemical phosphorus removal with filtration;
5. biological phosphorus removal; and
6. biological phosphorus removal with filtration.

For each of these treatment options, we used cost curves developed especially for this study to estimate costs for operation, maintenance, and sludge removal (OM&R), as well as capital costs to upgrade plants (Doran, 1997). Biological phosphorus removal is the least expensive option for point sources that currently have no treatment. For a 1 mgd plant, the cost of building a chemical phosphorus removal system is approximately $1.25 million. For the same size plant, a biological process would cost about $1 million. For both of these types of plants, costs per unit of treatment as measured by water flow decline significantly as the size of the plant increases. The costs of adding filtration to chemical or biological phosphorus removal processes is the same, about $2.5 million to capital costs for a 1 mgd plant.

Biological phosphorus removal processes have an advantage in total operation, maintenance and sludge removal costs, primarily because there is less sludge produced with biological processes. For a chemical process, the annual costs in this category are about $60,000 per mgd plant size

The watersheds of Saginaw Bay, the Minnesota River Valley, and the Rock River are quite similar in the dominance of agriculture in land use and phosphorus loads.

with no economy of scale. For biological removal, the comparable cost is $35,000 per year, with declining costs per unit of water treated as plants get larger. The additional OM&R costs for filtration for either process are about $80,000 per year, with a small decrease in unit costs as plants get bigger.[9]

Nonpoint Sources and Load Reduction Options

The watersheds of Saginaw Bay, the Minnesota River Valley, and the Rock River are quite similar in the dominance of agriculture in land use and phosphorus loads. As Table 6 shows, agriculture accounts for 47 to 76 percent of the total land use in each case. Pasture, urban, and other uses make up relatively small shares of land. Saginaw Bay has a heavily forested northern end, but in the other two cases forest land is not a large land use category. The Minnesota River Valley is the most rural of the three cases, with just 2 percent of the land in urban uses.

Corn for grain or silage is the dominant crop in each watershed, with almost 4 million acres in the Minnesota River watershed, 1 million in Saginaw Bay, and just over 600,000 acres in the Rock River. The Minnesota River Valley is the northern end of the Corn Belt, and the corn-soybean rotation is common there. Three million acres of soybeans were grown there in 1992. Soybeans are important in Saginaw Bay, with about 450,000 acres grown, but are much less important in the Rock River watershed, where pasture and hay are the second most important agricultural land uses. In all these cases, wheat, oats, barley, and other crops account for much smaller shares of the land used for agricultural production.

Phosphorus loads originate not only from point source discharges, but also from a variety of nonpoint sources such as agriculture, other

TABLE 6. **LAND USE FOR EACH WATERSHED** (PERCENTAGES IN PARENTHESES)

	Land Use (1,000 acres)				
	Cropland	**Pasture**	**Forestry**	**Urban**	**Total**
Minnesota River	7,636 (70)	692 (6)	279 (3)	231 (2)	10,842
Saginaw Bay	2,246 (40)	409 (7)	1,522 (27)	476 (9)	5,592
Rock River	1,413 (58)	203 (8)	159 (7)	235 (10)	2,420

Source: 1992 National Resources Inventory (USDA, 1994).

Note: Watershed totals include other land use categories not shown such as transportation, water, federal, minor and miscellaneous uses.

A principal mechanism by which phosphorus reaches surface waters is through erosion, as phosphorus attaches to soil particles and is carried away in surface runoff from the land.

land uses such as forestry, and urban runoff from golf courses, pets, and septic systems. A principal mechanism by which phosphorus reaches surface waters is through erosion, as phosphorus attaches to soil particles and is carried away in surface runoff from the land.

Agricultural land use is the major source of phosphorus for each of the watersheds included in this study. These results agree with EPA water quality assessments (USEPA, 1998). (*See Table 4.*) Analysis of erosion and expected phosphorus runoff from nonagricultural land (using standard technical coefficients) suggests that nonagricultural nonpoint sources are relatively minor contributors to the P load in these watersheds compared to other sources (Horner et al., 1994).

Because of agriculture's dominance in water quality problems, we used a more detailed methodology for estimating loadings from agricultural sources. This methodology was developed and peer reviewed as part of an earlier study done by the World Resources Institute to explore national policy options to enhance agriculture's economic and environmental performance (Faeth, 1995). The methods and data were adapted and extended to make them suitable for a watershed-level study.

To capture the impacts from agriculture, we looked at land use within the sub-basins of each watershed and developed alternatives for crop production systems commonly used within the region. Just as for point sources, the data collection was intended to describe the financial and environmental characteristics of cropping practices used in the watershed. This includes the principal crops grown, the rotations in which they are grown, and the tillage practices used.

The information used for this analysis came from a variety of USDA databases, including the Cropping Practices Survey, which is a state-level survey of farmers and the practices they use. From this information we identified all important production inputs and their levels of use for each crop rotation and tillage practice. Unfortunately, this source of information only covers the major field crops: corn, wheat, soybeans, hay, silage, barley, and oats. There is no similar source of production data for vegetables, sugar beets or other minor crops, so these are not included in the analysis. This is not a serious problem, however, because the crops covered account for 80 to 92 percent of cultivated cropland and erosion rates are similar for those crops included and excluded. To account for all loadings from agricultural land, we inflated the acreage of the crops included to equal the total for all cropland. From the 1992 National Resources Inventory, we have been able to identify the acreage of various cropping rotations and the acreage of each crop grown in each sub-basin (USDA, 1994b).

Tillage is critically important in determining the amount of soil that erodes from cropping activities, so we also represented five different tillage types:

- *mold board* plowing, which completely overturns the soil and leaves little or no residue to prevent erosion by wind or water;

Tillage is critically important in determining the amount of soil that erodes from cropping activities. About 15 percent of cropland is in some form of conservation tillage in the Rock River watershed, compared to 25 percent for the Minnesota River Valley and Saginaw Bay.

- *conventional tillage,* which uses a variety of implements such as disks and chisels but excludes the mold board and leaves some residue, but not much;
- *mulch tillage,* which leaves over 30 percent of the surface covered by residue to prevent runoff;
- *no-til,* which avoids tillage altogether, leaving all of the crop residue on the surface, cutting erosion by up to 90 percent; and
- *ridge tillage,* which uses built-up ridges within the field to intercept runoff and encourage percolation.

Tillage data were obtained from the Conservation Tillage Information Center county-level summaries and combined with the NRI crop rotation data to characterize the production alternatives available in each watershed. In total, we represent 10 rotations and four tillage types in 18 combinations.[10] Each of these combinations is also represented by the average nutrient use levels obtained from survey data and a nutrient management alternative that reduces fertilizer use to levels recommended by county extension agents.

The amount of conservation tillage, mulch, no-til, and ridge tillage varies for each watershed. About 15 percent of cropland is in some form of conservation tillage in the Rock River watershed, compared to 25 percent for the Minnesota River Valley and Saginaw Bay. The Rock River and Saginaw Bay have appreciably higher rates of mold board plow usage.

Environmental impact data such as soil erosion rates and nutrient loadings were estimated for each cropping alternative using a biophysical simulation model, EPIC, developed by the USDA (Williams et al., 1989). The results from the simulation model were calibrated to data from the county and sub-basin level. Crop yields and crop acreages were calibrated using county averages from the state agricultural statistics services and the 1992 U.S. Census of Agriculture (USDA, 1997b).

The value of damages caused by just sheet and rill erosion is $84 million per year in the Minnesota River Valley watershed, $13 million in Saginaw Bay, and $29 million in the Rock River.

Soil erosion rates for erosion by water (sheet and rill) and by wind were also calibrated so that the simulated totals for each sub-basin match totals estimated by the USDA's National Resources Inventory. Table 7 shows these erosion estimates, which use 1992 data. According to USDA, cropland soil erosion by wind is much larger than sheet and rill erosion for the Minnesota River Valley and Saginaw Bay. Values for the Rock River were not reported by the NRI.

USDA (Ribaudo, 1992) has estimated damages caused by sediment from sheet and rill erosion in 10 regions. Using a value for the

Great Lakes region of $5.07 per ton of soil eroded (1995 dollars), the value of damages caused by just sheet and rill erosion is $84 million per year in the Minnesota River Valley watershed, $13 million in Saginaw Bay, and $29 million in the Rock River.

Soil erosion is the principal means by which phosphorus from agricultural sources reaches rivers. However, not all soil erosion is a cause of environmental harm, because most eroded soil moves a short distance and does not reach a ditch, creek, or river where it can cause a problem. The Institute of Water Research at the Michigan State University (MSU) developed estimates for this study of the amount of sheet and rill erosion in each sub-basin that reaches a waterway (Ouyang and Bartholic, 1997). There are five methods for estimating these delivery ratios. MSU generated estimates using all five. We used the mean value of the five for the analysis. These "sediment delivery ratios" were used to estimate loads to the river by multiplying the acreage in a given practice by the amount of phosphorus attached to sediment for that practice. The range of delivery ratios for the sub-basins in each watershed are shown in Table 7.

Soil deposition by wind erosion is potentially a major source of phosphorus loading, given that wind erosion in the two cases where data exist is more than double sheet and rill erosion by water. Unfortunately, there is no method to estimate deposition of soil eroded by wind. For the Minnesota and Rock Rivers, we assumed a sediment delivery ratio for wind of 3 percent, which is approximately the area covered by water in these watersheds. For the Saginaw Bay, the prevailing winds flow from west to east over the bay, meaning that erosion from wind is likely to have a significant impact on the load. One study for a similar situation, the Chesapeake Bay, showed wind erosion deposition to the watershed to be 9 percent. We estimated that 10 percent of wind-eroded soils were deposited in the Saginaw watershed. Local soil conservationists believe this is a conservative estimate.

As noted earlier, animal production is also believed to be a large source of water contamination. Unfortunately, there are no good methods to estimate the environmental impacts of different forms of animal production or waste remediation options explicitly. In practice, manure from animal production is spread on agricultural fields as a nutrient source, and

TABLE 7. SOIL EROSION FROM CROPLAND AND DELIVERY RATIOS BY WATERSHED

Watershed	Sheet and Rill Erosion (1,000 tons per year)	Sediment Delivery Ratio for Sheet and Rill Erosion	Wind Erosion (1,000 tons per year)	Total Soil Erosion (1,000 tons per year)
Minnesota River	16,484	0.138 – 0.206	34,543	51,027
Saginaw Bay	2,640	0.171 – 0.216	6,702	9,342
Rock River	5,793	0.157 – 0.171	No estimate	5,793

Source: Erosion data, 1992 National Resources Inventory, USDA; delivery ratios, Ouyang and Bartholic, 1997.

erosion control, while not the only factor, is important in determining the ultimate disposition of the waste. Assuming that animal manure is all applied to cropland, then the load from animal operations should be reflected in the agricultural cropland load estimates.

In each of the three case studies, the phosphorus loads from agriculture are the dominant source, ranging from 51 to 69 percent of the total estimated load.

Total Load Estimates

Putting the point source and nonpoint source information together and simulating the baseline land use, we produced a baseload estimate for each watershed by sub-basin. The estimates by sub-basin vary considerably depending on the number and size and treatment of point sources and land use. The total and breakout by source category for each watershed is shown in Figure 5.

In each of the three case studies, the phosphorus loads from agriculture are the dominant source, ranging from 51 to 69 percent of the total estimated load. The total phosphorus load for the Minnesota River Valley was calibrated to estimates made by the MPCA. Since the point-source loads are reported by each facility, the adjustment was made to the nonpoint source loads. For Saginaw Bay, our results compare well to a prior study done by Limno-Tech (1983) that estimated total phosphorus loads to the Bay at 989 tons per year. For the Rock River, the approach we used compared well with per-acre load averages estimated by hydrological models.

FIGURE 5. PHOSPHORUS LOADS BY SOURCE AND WATERSHED

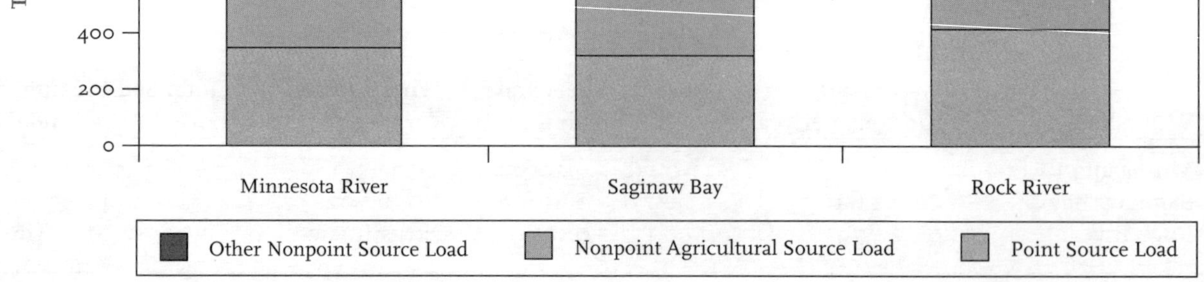

5

ECONOMIC RESULTS

Working with our partner organizations, we used a modeling framework developed for this study to analyze the economic and environmental potential of alternative policy approaches. Some of the assumptions were tailored to the data available in each watershed. The scenarios reported here are the same to allow cross-watershed comparisons. Since phosphorus is the principal pollutant of interest in each of these cases, we focus on phosphorus load reductions.

We asked the simple question, "What is the least-cost solution for a given level of phosphorus load reduction?"

THE LEAST-COST SOLUTION

In the first set of scenarios, we asked the simple question, "What is the least-cost solution for a given level of phosphorus load reduction?" The cost curves developed through this exercise provide a benchmark to measure the cost-effectiveness of other policy tests. From an economic point of view, the least-cost solution presents the ideal. The goal is to identify those policies that meet the environmental objectives and are as close as possible to the least-cost result.

Figure 6 shows least-cost curves for each watershed. These were obtained by running the model at total load reductions of 10, 33, 50 percent and at the maximum potentially attainable for each watershed. The maximum load reduction is different in each case because of the unique characteristics of each watershed. For each of these runs the model found the point source and cropping practice options that minimize the total cost to produce a load reduction of the specified amount; no policy constraints were applied.

As shown in Figure 6, significant load reductions are possible at very low costs. Costs increase slowly until a reduction of 50 percent is achieved, and then rise rapidly thereafter. The difference in costs is attributable to the fact that at lower levels of load reductions there are a number of relatively minor agricultural changes that can dramatically cut phosphorus loads. For example, for a cut of up to 10 percent, the principal shift would be from moldboard plowing to conventional tillage without the moldboard. The elimination of moldboard plowing would dramatically cut erosion—and therefore phosphorus loads—at a very low cost. At further load reductions, conventional tillage would be reduced and conservation tillage practices would become dominant. There would also be an increased move to nutrient management practices as greater reductions are demanded.

FIGURE 6. **LEAST-COST MODEL SOLUTIONS**

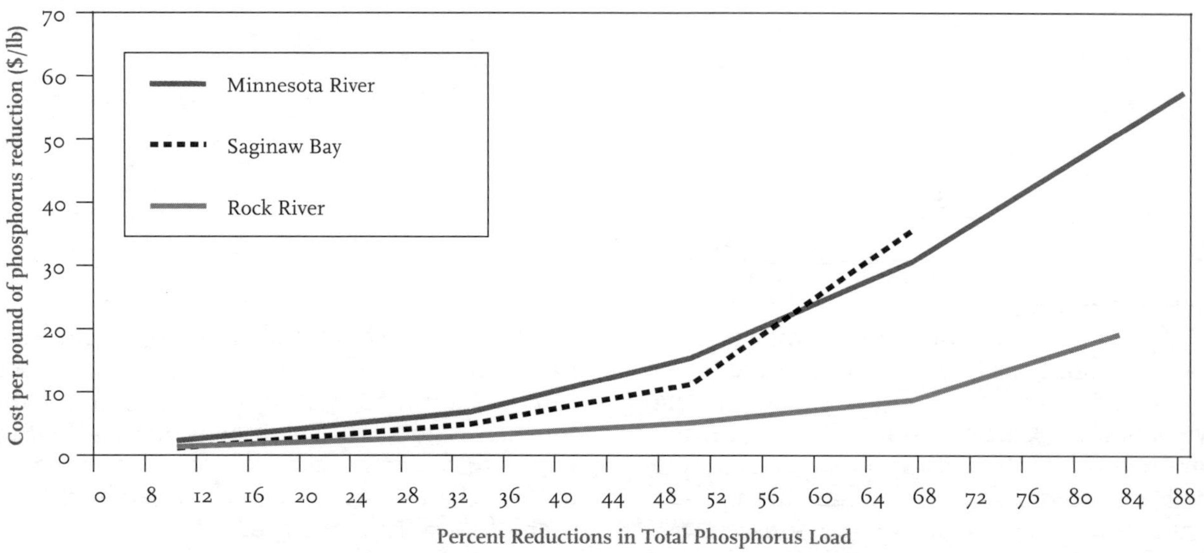

On the point source side, at relatively low reduction requirements, a few of the larger sources that have no systems treatment adopt biological or chemical phosphorus removal. As loads decrease, more plants adopt a removal process and those that have chemical removal reduce their loads by adding more of the chemical to the process, incurring only somewhat greater OM&R costs. Finally, at the maximum achievable reduction level, all point sources have adopted filtration, incurring the highest level of capital upgrade and greater OM&R costs. Reductions beyond the endpoint of the curves are not possible because at this point all point sources and agricultural options to reduce loads are employed to the fullest extent specified by the dataset. The opportunities for reductions are exhausted earlier for Saginaw Bay because a 1 ppm standard for phosphorus is already in place. The Minnesota River shows the highest eventual costs because of the predominance of small treatment facilities and the high unit costs of additional treatment.

POLICY TESTS

We tested several policy options that would typically be considered for situations like those presented by the three case studies. These were intended to contribute to the discussion on trading and were not endorsed by any state agencies. These tests include:

- *A point source performance requirement.* Point source controls have been the first avenue of attack to correct water quality problems. This scenario asks: "What if we do more of the same?" This scenario is a policy baseline for comparative purposes. We assumed in this scenario that all point sources would be forced to adopt a new standard for phosphorus, except in the case of Wisconsin, where the smallest dischargers would be exempted as a cost-saving measure. In Minnesota and Wisconsin, the standard would be 1 ppm. Michigan already has a standard of 1 ppm and would go to 0.5 ppm.

- *A conventional subsidy program for agricultural conservation "best management practices" (BMPs).* Instead of regulatory controls on agriculture, most of the policy effort in the U.S. has focused on providing subsidies to farmers to help reduce the costs of implementing best management practices (BMPs). In this vein, this scenario provides a subsidy for mulch tillage, no-til, and nutrient management. We adjusted the subsidy level (unique in each case) until it induced an agricultural load reduction for the watershed equal to that for the point source performance requirement.

- *A point source performance requirement coupled with trading.* What if point sources could trade with other point and nonpoint sources to meet the new standard? This scenario adds point and nonpoint trading flexibility to the first scenario.

- *A trading program coupled with performance-based conservation subsidies.* This scenario combines elements of the second and third scenarios, where the burden of reductions is shared between point source and nonpoint sources. A key difference however, would be that the conservation subsidies would be based upon the attainment of least-cost load reductions however achieved, not the adoption of any particular BMP. For nonpoint source reductions that are applied to nonpoint source obligations, there is no trading ratio applied.

In each case, the strategy of tightening point source performance requirements is the most expensive option.

Table 8 shows the cost results for all the case studies. Additional details for each case are shown in Tables 9a through 9c. In each case, the strategy of tightening point source performance requirements is the most expensive

TABLE 8. **COST OF PHOSPHORUS CONTROL UNDER DIFFERENT POLICIES (US$ PER POUND OF PHOSPHORUS REMOVED)**

	Point Source Performance Requirement	Conventional Subsidies for Agricultural Conservation Practices (BMPs)	Point Source Performance Requirement with Trading	Trading Program with Performance-based Conservation Subsidies	Least-cost Solution
Minnesota River	19.57	16.29	6.84	4.45	4.36
Saginaw Bay	23.89	5.76	4.04	2.90	1.75
Rock River	10.38	9.53	5.95	3.82	3.22

Source: The levels of phosphorus reduction from the base are different for each watershed.

TABLE 9a. SCENARIO RESULTS—MINNESOTA RIVER VALLEY

	Point Source Performance Requirement (1ppm for phosphorus)	Conventional Subsidies for Agricultural Conservation Practices (BMPs)	Point Source Performance Requirement with Trading	Trading Program with Performance-based Conservation Subsidies	Least-cost Solution
Total P Load Reduction (percent)	20	20	29	20	20
Point Source P Load Reduction (percent)	70	0	55	35	30
Agricultural P Load Reduction (percent)	0	30	20	15	17
Total Costs ($M/yr)	9.58	8.06	5.01	2.21	2.15
Average Cost per Pound ($/lb.P)	19.57	16.29	7.10	4.50	4.36

TABLE 9b. SCENARIO RESULTS—ROCK RIVER WATERSHED

	Point Source Performance Requirement (1ppm for phosphorus)	Conventional Subsidies for Agricultural Conservation Practices (BMPs)	Point Source Performance Requirement with Trading	Trading Program with Performance-based Conservation Subsidies	Least-cost Solution
Total P Load Reduction (percent)	30	70	40	30	30
Point Source P Load Reduction (percent)	71	0	59	35	24
Agricultural P Load Reduction (percent)	0	58	29	29	39
Total Costs ($M/yr)	5.9	5.4	4.5	2.2	1.8
Average Cost per Pound ($/lb.P)	10.38	9.53	5.95	3.82	3.22

TABLE 9c. **SCENARIO RESULTS—SAGINAW BAY**

	Point Source Performance Requirement (1ppm for phosphorus)	Conventional Subsidies for Agricultural Conservation Practices (BMPs)	Point Source Performance Requirement with Trading	Trading Program with Performance-based Conservation Subsidies	Least-cost Solution
Total P Load Reduction (percent)	16	16	27	18	16
Point Source P Load Reduction (percent)	49	0	36	21	0
Agricultural P Load Reduction (percent)	0	28	27	20	27
Total Costs ($M/yr)	7.47	1.80	2.14	1.06	0.55
Average Cost per Pound ($/lb.P)	23.89	5.76	4.04	2.90	1.75

option. Costs across the studies vary considerably, however. Saginaw Bay results are the highest because there is an existing requirement of 1 ppm in place and the requirement simulated here is for 0.5 ppm instead of 1 ppm as in the other studies. Minnesota River results are also relatively high because there are so many small sources. In contrast, in the Rock River the smallest sources are exempt and the costs are quite a bit lower for the same level of control.

Tables 9a-c show estimates for load reductions under each policy test. The new regulatory requirements on point sources provide a cut in the point source load of 70, 71, and 49 percent, respectively for Minnesota River, the Rock River, and Saginaw Bay. These reductions work out to cuts in the total load of 20, 30, and 16 percent. Again, less potential is available from point source reductions in Saginaw Bay because phosphorus is already controlled.

A conventional conservation subsidy program for agricultural BMPs is somewhat better at achieving the same result, but is still relatively expensive except in Saginaw Bay, where wind erosion is such a problem and soil conservation has a greater benefit. The costs of achieving the same amount of load reduction through untargeted agricultural subsidies for conservation tillage practices is lower than the point source regulation approach, but still more expensive than other options. Costs are comparatively greater in the Minnesota River because there is more adoption of conservation tillage in this watershed and wind erosion, unlike Saginaw, is less of a problem. There is also greater use of the moldboard plow, the most erosive practice, in the Rock River. These characteristics provide less expensive remediation opportunities.

If agriculture achieved the same absolute cuts as new point source performance requirements, the percentage load reduction would be less because the agricultural load in each case comprises a larger share of the total. We compared the cost per acre required in the model to bring in more conservation tillage with the costs actually paid under existing government programs. In each case, the results were quite close to the actual.

Point source performance requirements, coupled with flexibility to decide whether to treat or trade with point or nonpoint sources, would lower costs considerably.

Compared to the least-cost solution, both of these approaches are expensive. The least-cost solution relies on performance objectives to achieve the desired result and is otherwise unconstrained. The first two policy tests are obviously quite far from the most cost-effective result. In the case of new point source requirements, the reason is that there are many small point sources with diseconomies of scale and therefore expensive remediation costs compared to agriculture. The agricultural conservation subsidy program favors certain practices without regard to performance and is therefore relatively inefficient. The least-cost solution represents the equivalent of a highly targeted performance-based subsidy program. The least-cost solution ranges from 69 to 93 percent lower compared to tighter regulatory standards on point sources and from 66 to 73 percent lower than untargeted BMP subsidies.

In contrast, point source performance requirements, coupled with flexibility to decide whether to treat or trade with point or nonpoint sources, would lower costs considerably. In this case, the performance requirement acts as the "cap" in a "cap and trade" system. This would not represent a "fully closed" cap and trade program because not all sources are covered under the cap. Nevertheless, a point source standard with trading gets much closer to the least-cost solution because it allows the point sources to take advantage of the least expensive remediation opportunities, wherever they may be found. This scenario assumes that only one third of the load reduction from nonpoint sources could be applied against point source requirements (a 3:1 trading ratio). The rest of the load reduction produced is essentially an "environmental credit" to assure the achievement of water quality goals and account for the greater uncertainty inherent in nonpoint source loads. The trading ratio produces a greater total load reduction than any other policy, an additional 10 percent for each case. Even with the environmental credit for the uncertainty of the nonpoint source load applied to the point source obligations, there is still a significant savings over the strict regulatory case. For Saginaw Bay, the costs drop by nearly $20 per pound—82 percent for the point sources. The other case shows similar though less dramatic cost reductions, primarily because of the phosphorus requirement already in place in Michigan.

Is it fair to ask point sources to pay for remediation simply because that is where regulatory control is the strongest and politically easiest? Should broader social or sectoral responsibility be sought?

This trading scenario assumes that existing point sources would pick up the bill to help clean up threatened rivers, lakes, and estuaries, raising the question of equity. Is it fair to ask point sources to pay for remediation simply because that is where regulatory control is the strongest and politically easiest? Should broader social or sectoral responsibility be sought? These questions was raised in each of our local advisory panels.

In response to these issues, we constructed a scenario that mimics burden sharing. This scenario, "trading with performance-based conservation subsidies," assumes that the burden for reductions would be borne evenly by point and nonpoint sources. In this case, point sources would be responsible for half of the reduction level imposed in the other tests. The remaining reduction is assigned to the agricultural sector. We assumed for this scenario that the cost for agricultural reductions would come from conservation subsidy program funds. However, the subsidies would be performance-based, so that those farmers able to produce the cheapest load reductions would receive the available funds. This would mean that farmers closest to streams, with the most highly erodible soils, or who had not yet adopted conservation practices would be the first to receive program funds. Further, we assumed that point sources could still purchase nonpoint source reductions. Because agriculture has its own obligation, however, and would not trade away its cheapest reductions, the results show very little point-nonpoint source trading, but quite a bit of point-point source trading. This scenario effectively provides a cap for all sources and so is closer to a "fully closed" trading system.

Here, both point and nonpoint sources contribute evenly to the reduction of phosphorus loads. While most of the total costs would still be paid for by point sources, their costs would be cut considerably compared to the previous trading scenario and even more compared to a regulatory standard. For example, in the Rock River, the cost would be just $3.82 per pound. The total annual cost estimate to produce this reduction is about $2.2 million per year, less than half of the point source trading program by itself. Of this total cost, about $600,000 would be borne by farmers or by public subsidy programs. The total cut in the phosphorus load is less because the trading ratio is not applied to reductions made by agriculture on its own behalf, as in the case of the trading program with caps on point sources only. This scenario is the closest to achieving the economically ideal least-cost solution. Point sources still bear the majority of the costs, but these are the lowest of any scenario.

6

SUMMARY AND RECOMMENDATIONS

From the analysis presented here, it is clear that if water quality programs are designed more efficiently, significant water quality gains can be achieved at relatively low cost. A comparison of alternative programs show that there is a wide range in the cost-effectiveness of different approaches, with conventional strategies showing the least benefit per dollar spent. More flexible approaches can potentially provide greater improvements in water quality, over a larger range of reductions, and at much lower cost.

More flexible approaches can potentially provide greater improvements in water quality, over a larger range of reductions, and at much lower cost.

This is not to say that conventional regulatory approaches have been a failure in achieving improvements in water quality; they have not. But our analysis shows that pushing on point sources alone would be a relatively expensive approach when other sources contribute more to the problem. Conversely, in the agricultural sector the opportunities for inexpensive gains are great if conservation subsidies were to be based upon performance, to the extent that we are able to estimate it. Further, point sources are not the largest contributor to the problem in the watersheds we looked at, or on a national basis. This means that any strategy to reduce the level of loads to restore surface waters must include agriculture, not only because of economic efficiency arguments, but also for the sake of fairness. This may not be true in every watershed, but it appears to be a dominant situation in many rural and urbanizing U.S. watersheds.

Any strategy to reduce the level of loads to restore surface waters must include agriculture, not only because of economic efficiency arguments, but also for the sake of fairness.

Because there is a large differential between remediation costs for conventional approaches and programs that involve trading of some form, trading has potential in the watersheds we considered. One would also expect that there is potential for trading in many other watersheds as well, because the same reasons for cost-effectiveness would apply. And, as Table 3 shows, there are many watersheds in the United States that are impaired by nutrients.

While a regulatory mandate on point sources coupled with a flexible trading program appears to have merit, an even better program would couple these elements with a strategy directly involving agriculture. In such a program, point-point and point-nonpoint trading would be allowed, but nonpoint sources would have a shared responsibility to undertake remediation actions not coupled to point source regulatory requirements. This could take one of several forms. In one case, agricultural conservation subsidies, or some share of them, could become part of the pool of funds available through a joint trading program. Farmers, municipalities, or industrial sources who generated credits could sell them through a single program sponsored by the government or another broker in conjunction with point sources who wished to purchase and apply credits. Credits purchased by government conservation funds would be retired.

As a second option, farmers could only generate credits after they had met a minimum standard for agricultural practices. For example, a farmer who used a moldboard plow or applied manure to frozen fields in the wintertime could not generate credits by abandoning what are broadly acknowledged to be environmentally unsound practices. But, farmers moving from conventional tillage to conservation tillage, or farmers who currently had sound manure management practices but wished to install more advanced ones, could generate credits. Alternatively, farmers moving from practices that were below standard to ones well above standard could generate a partial credit. While such a program requirement would be entirely voluntary, it would have the benefit of rewarding farmers who undertake sound management on their own, and would still provide an incentive for farmers who had not yet made the change. It would have the disadvantage, however, of reducing cost-effectiveness.

Finally, state agencies could apply a performance requirement to all sources, including agriculture, and allow point sources and farmers the option of meeting the requirement through trading. Point and nonpoint sources would have access to the same pool of credits, and conservation subsidy funds could be applied as before to offset some part of farmers' costs or the cost of operating the program. Farmers who had previously undertaken conservation practices would be in compliance and have no further obligation. Others who could make inexpensive reductions would do so, and perhaps do more than their obligation to sell credits and recover their costs. Some farmers with high costs or high-valued crops who wished to continue their current practices could purchase credits from point or nonpoint sources to meet their obligation.

While trading has economic potential, there are some uncertainties associated with trading that need to be acknowledged and accounted for.

While trading has economic potential, there are some uncertainties associated with trading that need to be acknowledged and accounted for. The first and perhaps most important aspect of trading that would involve nonpoint sources is that there is a great deal of uncertainty involved because the loads are tied to weather events. While point sources produce fairly regular flows across seasons and even years, nonpoint sources do not. Loads are highest during rainy seasons and years with high precipitation, and conversely lower at other times. For this reason, a reduction in the load from a nonpoint source may not be equivalent to that from a point source. Therefore, it is important that water quality is monitored to make sure that expected improvements are realized and water quality goals are met.

A key advantage of the nonpoint source reductions, however, is that they have other water quality benefits in addition to the reduction in phosphorus loadings. For example, in the point source cap-with-trading scenario, at the same level of phosphorus reductions necessary to achieve the P load performance there would also be a one-third reduction in sediment loads and nitrogen loads for agriculture.

Issues regarding liability must also be carefully considered. When a point source fails to meet a legal requirement, the responsible regulatory agency has the ability to force the offender to comply. Any water quality program that employs trading must similarly provide a legal remedy for those instances when someone sells or applies a credit that has no environmental value. One way to do this is to have sellers voluntarily accept the regulatory obligations of the purchaser for that amount of the load. Alternatively, the buyer can make the seller a loan to be paid in credits. If the credits are not delivered on the specified date, the loan is due with interest. If the seller cannot provide the credits himself, he can purchase them on the market and turn them over.

Another consideration is that trading programs can be expensive to put in place and operate if poorly designed. Regulatory paperwork, information gathering and the process of identifying partners to trade with can create transaction costs that are prohibitive and make a trading program ineffectual. Administrative oversight needs to be sufficient to ensure good performance, but not so burdensome as to inhibit trading. Registration of trades should be efficient so that partners can easily hook up, report their trades, and get approval. When numerous nonpoint sources are involved, some sort of broker—for example, a cooperative—needs to be organized to coordinate the sale of credits and to verify them using standard techniques. For all these reasons, trading should occur within a regulatory program where rules and methods are standardized and appropriate review can be cost-effective, not permit by permit, which is expensive.

Trading should occur within a regulatory program where rules and methods are standardized and appropriate review can be cost-effective, not permit by permit, which is expensive.

RECOMMENDATIONS

The potential of market-based policies to produce cost savings and environmental progress appears large enough that much more effort should be focused on this front. The following recommendations could help further these objectives.

1. **The Environmental Protection Agency and state environmental agencies should allow trading among industrial and municipal point sources whenever new nutrient reduction requirements are put in place.** The issues with point-point trading are fairly straightforward and can be effectively managed. With the right set of standards, large cost savings can be achieved, as this study demonstrates. The standards developed should be written into regulations and ensure that the environmental results from a trade are the same, if not better than that expected under conventional regulations. Standards should only allow trading within

the watershed that defines the problem to be addressed and should also set trading ratios that encourage reductions upstream. Trading should only result in reduced loads and improved water quality. In cases where new requirements are put in place because of a water quality problem, trading ratios between point sources should be close to one. However, if a source wants to increase its emissions because of expansion or growth, higher trading ratios should apply so that a net gain in water quality results. Waste treatment facilities should not be able to use trading to stop using capacity already in place, but trading could be used to allow facilities to cover inadvertent and infrequent permit violations. Facilities that have generally poor records of environmental compliance should be prohibited from trading.

2. **EPA and the U.S. Department of Agriculture should provide much greater regulatory and financial support for state efforts to develop point-nonpoint source pilot trading programs and standards.** Pilot trading programs should be viewed as laboratories to learn how the potential of point-nonpoint source trading can be achieved. Such pilots, if properly supported, could provide cost savings and other benefits, such as sediment reductions, that can come from the implementation of better agricultural practices. Pilots in various settings should be supported through appropriate regulations and aggressively pursued with technical and financial resources from federal agencies. Federal and state agencies and other local stakeholders should participate as cooperative partners. EPA regional administrators should provide oversight of these programs, but allow sufficient flexibility to try different approaches and help find ways to make these programs work. Evaluation procedures should be built into pilot programs. If agreed goals are not met within a reasonable period, such as one permit period of five years, the pilots should be terminated and standard regulatory approaches applied in that situation.

3. **Concentrated animal feeding operations should be treated the same as industrial and municipal point sources, which means that they should face involuntary permitting processes and tight enforcement.** Currently, confined animal feeding operations smaller than 1,000 animal units are not regulated in most states under the Clean Water Act. This is roughly equivalent to a city of 25,000 to 50,000 people. Cities much smaller than this do not escape regulatory obligations, and neither should farmers. Given that domestic livestock produce 130 times the waste of humans in the United States, it is remarkable that only 5,000–10,000 of the 1.1 million farms with animals have to control their livestock wastes. With livestock waste trends going in the wrong direction, the nation will never be able to improve water quality if such a large source contributing such a great share of the problem is left uncontrolled. Minimum animal waste standards need to be put in place for all concentrated operations. These would eliminate the most egregious practices such as drainage and overflow of holding ponds directly into waterways, application of wastes onto frozen fields, animal defecation directly into streams, and overapplication of manure beyond the ability of crops to use the nutrients. Without exception, large concentrated animal operations should be treated like point sources and be subject to the same standards that apply to municipal and industrial facilities of the same size. They should receive effluent limitations and be required to adopt practices to meet these permit-based limits.

4. **Sufficient money from farm income subsidy programs should be shifted to conservation**

programs so that environmental problems caused by agriculture can be remediated using market-based approaches. In 1996, farm income support programs were changed so that payments were based on historical payments rather than production. Cutting the link between commodity production and income payments was a useful change that will encourage the adoption of cropping rotations and less monoculture, which is a good outcome environmentally. However, the next step needs to be taken—a much greater share of the billions of dollars spent on unrestricted farm income support should be tied to environmental remediation efforts so that polluted runoff from agriculture can be contained. Farmers and the nation should come to a compromise to continue support payments, but link the payments to environmental improvements. The programs should employ market-based mechanisms such as auctions, which ensure that the greatest environmental benefit is achieved for the money spent.

5. **Monitoring of water quality should be vastly increased.** A common comment regarding pilot trading programs is that monitoring in those watersheds should be increased so that environmental performance can be assessed. There is sound logic in increased monitoring to ascertain the results of trading pilots. However, this logic should not be restricted just to innovative programs, but should extend to all water quality programs. The scant evidence available suggests that the nation's water quality policies are not sufficient to make progress toward the goals of the Clean Water Act. Given the tens of billions of dollars that are spent on infrastructure and the hundreds of millions on conservation subsidies, it is penny-wise and pound-foolish to provide insufficient funding for water quality monitoring. With the current levels of funding for monitoring, we will never be able to adequately assess the effectiveness of these investments and make appropriate adjustments with confidence. There is a great deal of interest in water quality among the general public. Some states and localities have used this interest to develop volunteer monitoring programs.

6. **Federal and state agencies should explore opportunities for trading in large watersheds.** To date, trading has generally been considered in the context of relatively small watersheds entirely within state boundaries, such as the ones we examined in this study. However, there are several situations in the United States where trading could provide major cost savings in large basins. Two that merit such attention are the Chesapeake Bay and the Mississippi Basin. Drainage from the Mississippi is the cause of one of the nation's biggest water quality problems, the "dead zone" in the Gulf of Mexico caused by oxygen depletion from nutrient enrichment. Trading programs in large basins would present additional problems, mainly because trades would cross state boundaries and have to be managed at the federal rather than the state level. The allocation of loads and assignment of caps—something necessary to set up a trading program—would have to include major and minor watersheds and resemble a large TMDL process. The advantage however, would be that for the first time policy would be forced to account for large-scale performance objectives in a consistent manner that was ecologically relevant. The contributions of various programs to the final objective would have to be assessed as part of a larger framework instead of piecemeal, as is done now. This could do more to force improvements in water quality than conventional approaches now being considered.

CONCLUSION

The Clean Water Act has succeeded in controlling some sources of pollution, but with countervailing increases from other sources the result has been something akin to running in place. In many instances the nation has still not achieved the law's goals, and the job of cleaning up our waterways remains unfinished. One reason we have not moved to correct our water quality problems is that the conventional remedies are perceived as too expensive. Less expensive options involving other dischargers can produce significant improvements in water quality at low cost. Trading can provide a flexible alternative and encourage the use of least-cost options.

ABOUT THE AUTHOR

PAUL FAETH is Director of the Economics Program at the World Resources Institute. He currently directs several efforts: analysis of the cobenefits of climate policies and water quality; an assessment of policies to address the dead zone in the Gulf of Mexico; collaborative work with industry on climate change; the development of Internet-based trading for nutrients; ecosystem services and biodiversity protection; and an assessment of trade and its impact on the environment in Latin America. Faeth was WRI's Liaison to the Sustainable Agriculture Task Force of the President's Council on Sustainable Development organized by President Clinton. Previously, Faeth worked with the International Institute for Environment and Development, where he applied methods of systems analysis to examine development projects to improve their economic and environmental performance. He has also worked for the USDA's Economic Research Service on issues related to agricultural trade policy.

NOTES

1. The states and other jurisdictions do not use identical survey methods and criteria to rate their water quality. Furthermore, they may survey different waterbodies every two years. As a result, EPA cautions against comparing water quality information submitted during different time periods. Though EPA has discouraged aggregating surveys because of methodological differences, this is currently the only trend data available.

2. Data for 1970 to 1993 are from livestock and poultry inventories in the USDA Economics and Statistics System:

   Cattle—(http://usda.mannlib.cornell.edu/data-sets/livestock/95131/ct7084.exe, and /ct8595.exe)

   Chickens—Layers (http://usda.mannlib.cornell.edu/data-sets/livestock/89007/2/table006.wk1)

   Chickens–Broilers—(http://usda.mannlib.cornell.edu/data-sets/livestock/89007/3/table065.wk1)

   Hogs—(http://usda.mannlib.cornell.edu/data-sets/livestock/95131/hg4q6778.exe, /hg4q7987.exe, and hg4q8894.exe)

   Turkeys—(http://usda.mannlib.cornell.edu/data-sets/livestock/89007/5/table131.wk1).

   Data for 1994 to 1997 are from United States Department of Agriculture-National Agricultural Statistics Service, Agricultural Statistics 1998. Washington, DC: United States Government Printing Office, 1998. (http://www. usda.gov/nass/pubs/agr98/acro98.htm).

3. Manure production and nutrient content parameters for this analysis came from Lander et al., (1998) and USDA, 1998b.

4. An animal unit is defined as a cow with calf, or the equivalent of 1,000 pounds of liveweight.

5. There is a significant body of literature on emissions trading in general and nutrient trading in particular that largely describes the same small set of examples. For papers on emissions trading, see Hahn, 1988. For summaries of existing or potential nutrient trading activities, see Michigan DEQ, 1998, or USEPA, 1996.

6. See USEPA (1996) and the Michigan Department of Environmental Quality

website (1998a) for more complete lists of where nutrient trading is being considered.

7. The Michigan DEQ prefers to use the term "water quality trading" rather than nutrient trading, because the explicit intent of their program is to use trading to improve water quality beyond regulatory requirements.

8. This average is weighted by flow. Only the two largest facilities have some phosphorus control, so the majority of plants have a considerably higher concentration of phosphorus in their effluent, although their share of the flow is small.

9. The cost curves developed by Doran (1997) are as follows (where Q is flow in mgd). For capital costs (in millions of dollars): standard chemical treatment $(1.25*Q)^{0.67}$; biological treatment $(1.0*Q)^{0.67}$; for filtration add $(2.5*Q)^{0.67}$. For operation and maintenance (in millions of dollars): standard chemical 0.6*Q; biological treatment $(0.035*Q)^{0.7}$; for filtration add $(0.08*Q)^{0.9}$.

10. Not all rotations employ each tillage practice so there are not 40 combinations, which might be expected. For example, the soybean-wheat rotation is predominantly grown using conventional tillage only.

REFERENCES

Adler, Robert W., Jessica C. Landman, and Diane Cameron. 1993. *The Clean Water Act 20 Years Later*. Natural Resources Defense Council. Washington, DC: Island Press.

Arbuckle, J. Gordon et al. 1993. *Environmental Law Handbook, Twelfth Edition*, Rockville, MD: Government Institutes, Inc.

Association of Metropolitan Sewerage Agencies (AMSA) and the Water Environment Federation (WEF). *The Cost of Clean*. Washington, DC, March 1999.

Bacon, Elise F. *Use of Economic Instruments for Water Pollution Control: Applicability of Point Source/Nonpoint Source Trading for Pollutant Discharge Reductions to Washington State*. Apogee Research, Inc. Bethesda, MD: September, 1992.

Batchelor, David. Water Quality Program Trading Specialist, Michigan Department of Environmental Quality, Lansing, MI. Personal communication, May 6, 1999.

Bosch, Darrell, Keith O. Fuglie, and Russ W. Keim. 1994. Economic and Environmental Effects of Testing for Fertilizer Management. Staff Report. Economic Research Service, USDA. Washington, DC.

Bosch, Darrell and Krishna Napit. 1992. "Economics of Transporting Poultry Litter to Achieve More Effective Use as Fertilizer." *Journal of Soil and Water Conservation* 47(4): 335–347.

Carpenter, S.R., N.F. Caraco, D.L. Correll, R.W. Howarth, A.N. Sharpley, and V.H. Smith. 1998. *Nonpoint Pollution of Surface Waters with Phosphorus and Nitrogen, Issues in Ecology*. Washington, DC: Ecological Society of America.

Doran, Michael D. *POTW Phosphorus Control Costs*. Unpublished memorandum. April 27, 1997. Madison, WI: Strand Associates.

Environmental Law Institute. 1997. *Enforceable State Mechanisms for the Control of Nonpoint Source Water Pollution*.

Faeth, Paul. 1995. *Growing Green: Enhancing the Economic and Environmental Performance of U.S. Agriculture*. Washington, DC: World Resources Institute.

Hahn, Robert W. and Gordon L. Hester. "Marketable Permits: Lessons for Theory and Practice." *Ecology Law Quarterly* Vol. 16:361.

Hoag, Dana L. and Jennie S. Hughes-Popp. 1997. "Theory and Practice of Pollution Credit Trading in Water Quality Management." *Review of Agricultural Economics*. Volume 19, Number 2, pp. 252–262.

Horner, Richard et al. *Fundamentals of Urban Runoff Management: Technical and Institutional Issues*. Washington, DC: The Terrene Institute. August, 1994.

Industrial Economics, Incorporated. 1987. *Point and non-point source pollution reduction trading along the Lower Minnesota River: Project Planning Briefing*. Unpublished paper prepared for the Regulatory Innovations Staff, Office of Policy, Planning, and Evaluation, USEPA. November 16, 1987.

Lander, Charles H., David Moffitt and Klaus Alt. 1998. *Nutrients Available from Livestock Manure Relative to Crop Growth Requirements*, Resource Assessment and Strategic Planning Working Paper 98-1. Washington, DC: U.S. Department of Agriculture, Natural Resources Conservation Service. *http://www.nhq.nrcs.usda.gov/land/pubs/nlapp1b.html*.

Limno-Tech, Inc. *Cost-Effectiveness of Conservation Tillage Practices for Improving Saginaw Bay Water Quality*. Ann Arbor, MI. June 1983.

Metcalf and Eddy in association with Bonestroo, Rosene, Anderik, and Associates, Inc., Dahlgren, Shardlow, and Urban, Inc., Poquette and Associates, and Martinez Corp. 1993. *Centralization/Decentralization Planning Study*. Draft Final Report. Volume 1. MWCC Project No. 920200. Metroplitan Waste Control Commission, Twin Cities Area.

Michigan Department of Environmental Quality (DEQ). 1998a. *Great Lakes Trading Network, Second Semi-annual Report to the Great Lakes Protection Fund*. Water Quality Trading—Home Page. Surface Water Quality Division. *http://www.deq.state.mi.us/swq/*.

Michigan DEQ. 1998b. *Part XXX. Water Quality Trading—Draft #9, 11/98*. Lansing, MI: Department of Environmental Quality, Surface Water Quality Division, Water Quality Trading Workgroup Discussion Document. *http://www.deq.state.mi.us/swq/*.

Minnesota Agricultural Statistics Service. 1996. *Minnesota Agricultural Statistics, 1996*. St. Paul, MN: Minnesota Department of Agriculture.

Minnesota Pollution Control Agency (MPCA). 1997a. *Governor Carlson Lauds Innovative Pollution Controls for Rahr Malting Company*. Press Release, January 28, 1997.

——. 1997b. *National Pollutant Discharge Elimination System and State Disposal System Permit MN 0031917*. St. Paul, MN: State of Minnesota, Minnesota Pollution Control Agency, Water Quality Division. January 8, 1997.

——. 1997c. Point-Nonpoint Source Trading Summary: Finalized for Rahr Malting Permit on January 8, 1997. St. Paul, MN: State of Minnesota, Minnesota Pollution Control Agency, Water Quality Division.

——. 1997d. Nonpoint Source Trade Crediting Calculation: Finalized for Rahr Malting Permit on January 8, 1997. St. Paul, MN: State of Minnesota, Minnesota Pollution Control Agency, Water Quality Division.

National Research Council, Committee on Long-Range Soil and Water Conservation, Board on Agriculture. 1993. *Soil and Water Quality: An Agenda for Agriculture*. Washington, DC: National Academy Press.

Ouyang, Da and Jon Bartholic. 1997. *Estimating Sediment Delivery Ratios for Three Midwestern Drainage Basins*. Unpublished white paper prepared for the World Resources Institute by the Michigan State Water Resources Lab.

Ribaudo, Marc. 1992. *Water Quality Benefits from the Conservation Reserve Program*. Agricultural Economic Report Number 606. Washington, DC. Resources and Technology Division. Economic Research Service, USDA.

Stephenson, Kurt and Leonard Shabman. 1996. Effluent Allowance Trading: A New Approach to Watershed Management. *Water Science Reporter*. Blacksburg, VA: Virginia Water Resources Research Lab, Virginia Tech, for the National Institute for Water Resources.

United States Department of Agriculture (USDA). 1994a. *Summary Report. 1992 National Resources Inventory*. Washington, DC: Soil Conservation Service.

——. 1994b. *1992 National Resources Inventory, CD #3 Midwest Region*. Fort Worth: SCS National Cartography and GIS Center.

——. 1997a. *Agricultural Resources and Environmental Indicators, 1996–97*. Washington, DC: Economic Research Service, Natural Resources and Environment Division. Agricultural Handbook No. 712.

——. 1997b. United States Agricultural Census. Information obtained from *http://www.nass.usda.gov/census/* for general information and *http://govinfo.library.orst.edu/cgi-bin/ag-state?*Minnesota (or Michigan or Wisconsin) to find county by county data for Minnesota. Information downloaded during 1997.

——. 1998a *Agricultural Statistics* (various years). National Agricultural Statistics Service. *http://www.usda.gov/nass/pubs/agstats.htm* *http://www.usda.gov/nass/pubs/agr98/acro98.htm*

——. 1998b. National Engineering Handbook Part 651: Agricultural Waste Management Field Handbook. Washington, DC: Natural Resources Conservation Service *(http://www.ncg.nrcs.usda.gov/tech_ref.html)*

United States Department of Commerce. 1998. *Statistical Abstract of the United States: 1998*. 118th Edition. Washington, DC.

United States Environmental Protection Agency (USEPA). 1980. *1980 Needs Survey: Cost Estimates of Construction of Publicly-Owned Wastewater Treatment Facilities*. EPA430/9-81-001. Washington, DC: Office of Wastewater Management.

——. 1992 Report to Congress.

——. 1992a. *Incentive Analysis for Clean Water Act Reauthorization: Point/Nonpoint Source Trading for Nutrient Discharge Reductions*. Washington, DC: Office of Water and Office of Policy, Planning and Evaluation.

——. 1992b. *TMDL Case Study: The Lower Minnesota River*. EPA841-F-92-011. Washington DC: Office of Water (WH-553).

——. 1994. President Clinton's Clean Water Initiative: Analysis of Costs and Benefits. EPA-800-R-94-002. Washington, DC: Office of Water.

——. 1996. Draft Framework for Watershed-Based Trading. Washington, DC: Office of Water. EPA-800-R-96-001.

——. 1997. Index of watershed indicators. *http://www.epa.gov/surf/hucinfo*. Visited on October 31, 1997.

——. 1999. 1998 TMDL Tracking System Data Version 1.1.

http://www.epa.gov/owow/tmdl/trcksys.html. Visited on July 12, 1999.

United States Senate. 1997. *Animal Waste Pollution in America: Environmental Risks of Livestock and Poultry Production.* Report Compiled by the Minority Staff of the United States Senate Committee on Agriculture, Nutrition, and Forestry for Senator Tom Harkin (D-IA), Ranking Member. Washington, DC.

Williams, J.R. et al. 1989. *EPIC—Erosion/ Productivity Impact Calculator: 2, User Manual.* Technical Bulletin No. 1768. A.N. Sharpley and J.R. Williams, eds. Washington, DC: USDA.

ADDITIONAL REFERENCES

Anderson, 1999 *(see EPA website)*

World Resources Institute

The World Resources Institute (WRI) is an independent center for policy research and technical assistance on global environmental and development issues. WRI's mission is to move human society to live in ways that protect Earth's environment and its capacity to provide for the needs and aspirations of current and future generations.

Because people are inspired by ideas, empowered by knowledge, and moved to change by greater understanding, the Institute provides—and helps other institutions provide—objective information and practical proposals for policy and institutional change that will foster environmentally sound, socially equitable development. WRI's particular concerns are with globally significant environmental problems and their interaction with economic development and social equity at all levels.

The Institute's current areas of work include economics, forests, biodiversity, climate change, energy, sustainable agriculture, resource and environmental information, trade, technology, national strategies for environmental and resource management, and business liaison.

In all of its policy research and work with institutions, WRI tries to build bridges between ideas and action, meshing the insights of scientific research, economic and institutional analyses, and practical experience with the need for open and participatory decision-making.

WORLD RESOURCES INSTITUTE
10 G Street, N.E.
Washington, D.C. 20002, USA
http://www.wri.org/wri